THE COMPLE

BAOFENG RADIO BIBLE

FOR BEGINNERS

FAST-TRACK YOUR WAY TO BECOMING A BAOFENG PRO WITH AN EASY-TO-FOLLOW GUIDE TO STAYING CONNECTED WHENEVER IT MASTER MOST

BY

ELIAB BARNET

© Copyright 2024 by **Eliab Barnet & EchoWords Publishing**

- All rights reserved.

This document is geared towards providing exact and reliable information regarding the topic and issue covered.

- From a Declaration of Principles accepted and approved equally by a Committee of the American Bar Association and a Committee of Publishers and Associations.

It is not legal to reproduce, duplicate, or transmit any part of this document electronically or in printed format. All rights reserved.

The information provided herein is truthful and consistent in that any liability, in terms of inattention or otherwise, by any usage or abuse of any policies, processes, or directions contained within is the solitary and utter responsibility of the recipient reader. Under no circumstances will any legal responsibility or blame be held against the publisher for any reparation, damages, or monetary loss due to the information herein, either directly or indirectly.

Respective authors own all copyrights not held by the publisher.

The information herein is solely offered for informational purposes and is universal. The presentation of the information is without a contract or any guarantee assurance.

The trademarks used are without any consent, and the trademark publication is without permission or backing by the trademark owner. All trademarks and brands within this book are for clarifying purposes only and owned by the owners, not affiliated with this document.

⭐ HERE IS YOUR FREE GIFT! ⭐

👇 SCAN HERE TO DOWNLOAD IT 👇

TABLE OF CONTENTS

INTRODUCTION   9

1. THE COMPLETE BAOFENG OVERVIEW   11

1.1 HISTORY AND EVOLUTION   11
1.2 COMPARING MODELS   12
   HANDHELD TRANSCEIVERS   12
   MOBILE/BASE STATIONS   13
   DUAL-BAND RADIOS   13
   WATERPROOF AND WEATHERPROOF RADIOS   13
   FEATURE COMPARISON   14
   PERFORMANCE COMPARISON   15
   SUITABILITY COMPARISON   16
1.3 TECHNOLOGICAL ADVANCES   17
   TECHNOLOGICAL INNOVATIONS IN BAOFENG RADIOS: 17

2. BEYOND THE BASICS - ADVANCED OPERATIONS   19

2.1 PROGRAMMING MASTERY   19
2.2 CYBERSECURITY FOR RADIO COMMUNICATIONS: ENCRYPTION AND SIGNAL PROTECTION   21
2.3 INNOVATIONS IN RADIO USE   23

3. EXPLORING ACCESSORIES AND ADD-ONS   25

3.1 ANTENNAS   27
3.2 BATTERIES AND CHARGERS   29
3.3 CASES AND HOLSTERS   30
3.4 HEADSETS AND MICROPHONES   32
3.5 ACCESSORIES MAINTENANCE   33
3.6 USER TESTIMONIALS   35
3.7 CASE STUDIES   36

4. IN-DEPTH BAOFENG CUSTOMIZATION   37

4.1 TAILORED CONFIGURATIONS: HOW TO CUSTOMIZE RADIOS FOR PERSONAL NEEDS   37
4.2 ANTENNA SCIENCE: ANTENNA SELECTION AND TUNING FOR OPTIMAL PERFORMANCE   39
4.3 BATTERY AND POWER MANAGEMENT   41

5. LEGAL INSIGHT AND GLOBAL COMPLIANCE — 43

5.1 Navigating Regulations — 43
5. 2 Ethical Practices: Advocating Responsible Communication Ethics — 45
5.3 International Use: Tips for Travelers Using Baofeng Across Borders — 47
5.4 Case Studies and Examples: — 49
5.6 Obtaining Licenses — 50
5.7 Professional Insights — 52

6. REAL-WORLD APPLICATIONS AND CASE STUDIES — 53

6.1 Emergency Preparedness — 53
6.1.1 Creation of Strong Communication Strategies — 53
6.1.2. Testimonials and Real-Life Examples — 54
6.2 Baofeng in Professional Settings — 56
6.2.1. Security and Event Management — 56
6.2.2 Industrial and Commercial Applications — 57
6.3 Community Engagement — 58
6.3.1 Leveraging Online Platforms — 58
6.3.2 Educational and Outreach Programs — 58

7. ADVANCED COMMUNICATION STRATEGIES — 60

7.1 Digital Age Radio — 60
7.1.1 An Integration to Digital Technologies — 60
7.2 Covert Communications — 62
7.2.1 Techniques for Secure Communications — 62
7.2.3. Covert Operations Case Studies — 63
7.3 Long-Range Communication Tactics — 64
7.3.1 Maximizing Radio Reach — 64
7.3.2. Expert Tips and Techniques — 64
7.4 Case Studies — 65
Communication Standards in Responses to Emergencies. — 65
Covert Communication in Law Enforcement Activity Operations — 65
7.5 How to create a secure communication Network — 66
Choose Secure Communication Protocols — 66
Frequency Hopping Spread Spectrum (FHSS) Setup — 66
Implement Authentication Mechanisms — 66
Create Secure Communication Channels — 66
Test and Validate Network Security — 66
Monitor and Manage Network Traffic — 67
7.6 Troubleshooting Tips for creating and maintaining a secure connection network — 68
Signal Interference — 68

AUTHENTICATION FAILURES	68
NETWORK CONGESTION	68
ENCRYPTION KEY MANAGEMENT	68
HARDWARE MALFUNCTIONS	69
SECURITY BREACHES	69
ENVIRONMENTAL FACTORS	69

SPECIAL CONTENT 1: EMERGENCY COMMUNICATION CHECKLIST — 70

SPECIAL CONTENT 2: THE ULTIMATE HAM RADIO GUIDE — 71

8. THE TECHNICAL DEEP DIVE — 72

8.1 SIGNAL THEORY AND COMMUNICATION	**72**
8.1.1 BASICS OF RADIO WAVES	72
FACTORS AFFECTING PROPAGATION	73
8.1.2 PRACTICAL APPLICATIONS	74
Signal Strength Measurement	74
Antenna Selection and Positioning	75
Interference Mitigation	75
8.2 ANTENNA THEORY	**77**
8.2.1 CUSTOMIZING THE ANTENNA FOR SPECIFIC PURPOSES	77
Directional antennas	77
Omnidirectional antennas	78
Design Phase	79
Construction Phase	79
Tuning and Testing Phase:	79
Optimization Phase:	80
Troubleshooting Common Signal Issues	80
Signal Fading:	81
8.3 TECHNICAL TROUBLESHOOTING GUIDE	**83**
Common Issues:	83
Poor Reception:	84
Diagnostic Tools	84
SWR Meter:	84
Spectrum Analyzer:	85
8.3.2 MAINTENANCE AND REPAIR	85
Routine Maintenance	85
Firmware Updates	86
Simple Repairs	86
Speaker Replacement	86
Button or Knob Repair	87
8.4 SOFTWARE AND FIRMWARE UPDATES	**88**

8.4.1 KEEPING YOUR RADIO UPDATED	88
Importance of Updates	88
Performing Updates	88
Update Procedure	88
8.3.2 CUSTOMIZING SOFTWARE FEATURES	88
Third-Party Software	88
User Settings	88
Display Settings:	88
8.5 EXPERT INSIGHTS	**90**

9. INTEGRATING BAOFENG WITH MODERN TECHNOLOGY — 91

9.1 BRIDGING TECHNOLOGIES	**91**
COMPARATIVE ANALYSIS OF INTEGRATION TECHNOLOGIES	91
Bluetooth Integration:	92
Wi-Fi Integration:	92
Cellular Connectivity:	92
PRACTICAL USE CASES	93
INTERACT WITH THE INTERNET OF THINGS (IOT) PLATFORMS	93
Agricultural Monitoring Systems:	93
Asset Tracking Applications:	94
Smart Home Automation Projects:	94
9.2 REMOTE CONTROL APPLICATIONS	**94**
TECHNICAL SETUP AND CONSIDERATIONS	95
DEVELOPMENTS IN THIS DISTANT CONNECTION	97
9.3 DATA TRANSMISSION	**98**
METHODS AND PROTOCOLS	98
Security Aspects:	99
Case Studies:	99

10. BUILDING YOUR GROUND STATION — 100

10.1 DESIGN AND SETUP	**100**
TROUBLESHOOTING AND OPTIMIZATION	101
10.2 NETWORKING RADIOS	**102**
CREATING RADIO NETWORKS	102
NETWORK MANAGEMENT AND PROTECTION	103
10.3 ADVANCED SCANNING TECHNIQUES	**104**
DEVELOPING SCANNING SKILLS	104
EMBEDDING SCANNING INTO DAILY OPERATIONS.	104
10.4 INTERACTIVE EXAMPLE	**105**
SCENARIO: COMMUNITY EVENT COMMUNICATION STATIONS.	105
10.5 DIGITAL INTEGRATION	**107**

DIGITAL VOICE MODES	107
REPEATER CONTROLLERS	107
NETWORK MONITORING TOOLS	107
SOFTWARE SOLUTIONS	108
10.6 SECURITY CONSIDERATIONS	**109**
ENCRYPTION PROTOCOLS	109
ACCESS CONTROL MECHANISMS	109
NETWORK SEGMENTATION	109
SECURE AUTHENTICATION METHODS	109

CONCLUSION 111

CONTINUOUS LEARNING	111
PREDICTING FUTURE TRENDS	112

GLOSSARY 113

REFERENCES 125

INTRODUCTION

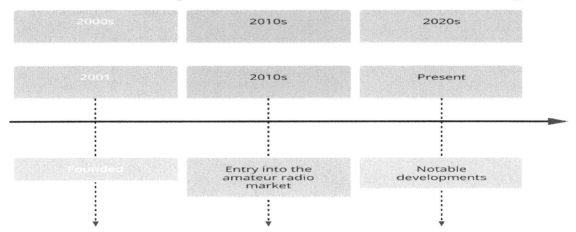

The Baofeng effect is a new development in amateur radio and other communication technology, astonishing the minds of hobbyists and newbies alike with its remarkable multi-faceted character and accessibility. Since its inception in 2001 in China, Baofeng Corporation has produced two-way radios for various commercial and industrial applications. It was not until the 2010s that Baofeng became known in the amateur radio community by releasing cheap and accessible ham radio models for many people who had not previously managed it.

Baofeng has become synonymous with amateur radio technology among radio enthusiasts, preppers, and tech-savvy people who love the idea that anyone can access this technology. Baofeng now encompasses several models at budget-friendly prices, thus enabling people to participate in radio communication and enjoy it more freely. Whether you are a pro ham radio operator or just a newbie enthusiast, Baofeng Radio will have an appropriate radio model.

Nevertheless, despite the acknowledged fame of Baofeng, it cannot entirely escape the uproar and disagreement. Some purists believe that Baofeng radios are a source of degradation of a standard operating procedure in amateur radio practice. In contrast, others find them an excellent tool for opening up to the hobby. Furthermore, these issues include discussion of the build quality and reliability and the illegal and non-compliance of the regulatory process.

For newcomers, navigating the vast and varied ecosystem of accessories, software, and communities can take time and effort. From programming cables and antennas to third-party software and online forums, plenty of resources are available to Baofeng enthusiasts seeking to enhance their radio experience. However, it requires diligence and critical thinking to discern reliable information from the noise of online discourse.

The allure and impact of Baofeng radios extend far beyond their technical specifications. They represent a cultural shift in how we communicate, prepare for emergencies, and connect with others in an increasingly interconnected world. Baofeng radios have undoubtedly left an indelible mark on amateur radio and communication technology.

 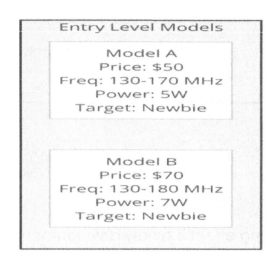

In the following chapters, we'll delve deeper into the rich history, diverse features, and practical applications of Baofeng radios.

Over the long run, Baofeng gained many accolades and some arguments for its gear. Confronted with events, Baofeng responded positively and indicated more steps to be creditworthy. Through conducting user feedback surveys and working on solutions to customer problems, Baofeng proves its dedication to improving and meeting the needs of its customers. Here are a few user testimonials:

George S.:

"Through the years since I started using Baofeng radios, I've been amazed at the efficiency and resilience each time. It doesn't matter whether I am high up there at the mountains hiking or I have volunteered to assist in emergency rescue efforts, my Baofeng device is always improved my connection when it really matters."

Kara M.:

"As a professional event coordination company, communication is key to ensuring our events will thrive. Walkie-Talkies have been a great asset to us, delivering clear audio and reliable coverage even in choppy venues. We do not trust any other brand for our events."

Whether you're a seasoned enthusiast or a curious newcomer, we'll offer insights and guidance to help you explore the boundless possibilities of amateur radio communication in the 21st century. Join us as we unravel the mysteries of the Baofeng phenomenon.

1. THE COMPLETE BAOFENG OVERVIEW

1.1 HISTORY AND EVOLUTION

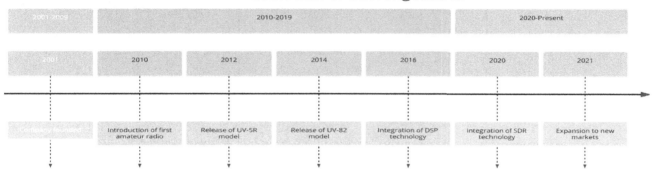

Evolution of Baofeng Radios

Baofeng Corporation is a Chinese company operating since 2001. The organization's beginning saw the focus on developing two-way radios, which were meant for commercial and industrial use. The amateur market did not begin to feel the influence of this company until 2010, when the company released its first amateur radio. Baofeng grabbed the attention of many customers at CES with the release of its amateur radio product line. This move represents a new stage of development for the company. These radios provided an affordable price to users and enjoyed a clever feature set that proliferated and won fans' loyalty globally.

The Baofeng guidance focused on addressing the supply gap for these radios, and that demand drove the company to increase product diversity to cover a wide range of users. However, they started to manufacture radios with an aim that could be favorable to both pro ham radio users and casual hobbyists. The migration of Baofeng to the amateur radio industry has never been just a move to an already established niche. It is a step to lead the evolution of amateur radio applications. The firm has been at the cutting edge of research & development in feature spectrum and bit capacity enhancement. In addition to these technologies, they have also improved the algorithms for digital signal processing in their radios – another consideration contributing to the radio's sensitivity and reliability.

Also, Baofeng's walk from humble roots to the global community's threshold proves that inventiveness and endurance can make great fortunes. The company's victories are the fruit of its ability to conform to the market's demands and maintain the ethical standards of offering quality and affordable products without compromising reliability and performance. Today, Baofeng Corporation is the leading company in the amateur radio market, and hence, Baofeng radios are used and admired by enthusiasts worldwide.

1.2 COMPARING MODELS

To assist users in selecting the right Baofeng radio for their needs, let's compare some key models: ~~To assist users in choosing the right Baofeng radio for their needs, let's compare some key models:~~

- **Baofeng UV-5R**: It is one of the handheld transceivers with an audition for being affordable and hoards many accessories and versatility with its dual-band operating capacity.

    I have (margin note)

    VHF (2-m band)
    UHF (70-cm band)

Baofeng BF-F8HP: A high-power handheld transceiver capable of all conditions, with higher transmit power and more battery capacity, and is used for long-distance communication.

Baofeng UV-50X3: A dual-band car/fixed station with a compact design featuring long-range features, the most effective one for both vehicle-installed and fixed base stations.

HANDHELD TRANSCEIVERS

Handheld transceivers are small, lightweight, and travel-convenient radios tailored for transmitting and receiving messages during movement. Radio and other emergency operators praise moderately priced devices such as the Baofeng UV-5R and UV-82 for their affordability and flexibility. At the same time, handheld transceivers have been found to offload the challenges of persons who have to move without losing their communication capabilities.

MOBILE/BASE STATIONS

Stations for mobile and base are purposely made for maximum signal output during vehicle use. This allows them to travel longer ranges compared to handheld transceivers. Like the Baofeng BF-F8HP model and others on that line, these models are ideal for installation in a vehicle or a fixed base station (one stationary) for home/office installations. This radio overcomes the transmission challenge, making it a reliable option for users who need a stable link beyond the average distance.

DUAL-BAND RADIOS

The two frequency bands of these radios can be used individually or within the Dual-band radios for VHF and UHF operations. Channels do not limit these radios; they provide users greater flexibility and coverage while allowing communication across different frequency ranges. Sources: American Studies, Issue 3: Imagining Home The Baofeng UV-5R and UV-82 represent two widely used dual-band radios. Such radios are ideal because they provide the most convenient communication and ensure safety in many situations.

WATERPROOF AND WEATHERPROOF RADIOS

Baofeng also produces dust & waterproof radios explicitly designed for the wearers to use in extreme places. These radios are built to be strict with water, dust, and impact-combined resistance, thus suitable for camping, hiking, and emergencies. Equipped with models that allow them to function in unfavorable weather situations, these radios can give their users a trustworthy source of communication during trips with complex environments.

FEATURE COMPARISON

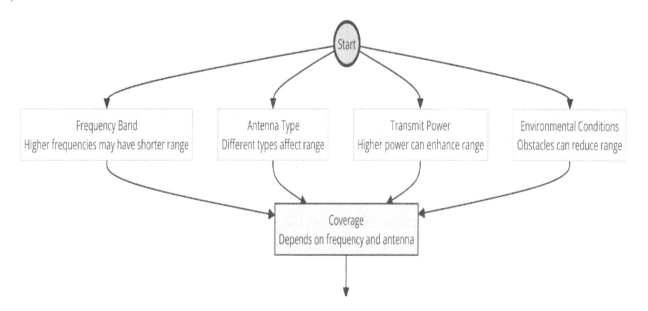

To compare Baofeng radio models effectively, evaluating their key features and specifications across various categories is essential. To compare Baofeng radio models effectively, it's necessary to assess their key features and specifications across multiple categories:

Frequency Range: The options for Baofeng radios cover the vast spectrum of VHF frequencies (Very High Frequency) and UHF bands (Ultra High Frequency). This determines the coverage area and possible interference with other radio facilities.

Power Output: STS: One of the crucial parts of transmission is the power output, which denotes the strength of the radio signal emitted by the device, usually measured in watts. Overall, higher power produced yields both extended range and allows the signal to penetrate obstacles.

Channel Capacity: The number of radio channels available for a channel depends on the communication flexibility and variability. Radio with a more significant channel can register more frequencies and several communication channels. This is a welcomed addition as it allows for more organized and efficient dealings among communication channels.

Display and Interface: The interrelation between the radio's and its interface's design and interaction are the key to user interaction and ease of use. Models with huge, backlit transceivers and easy-to-operate user interfaces provide more light and simplicity, essential in poor lights and outdoor spaces.

Battery Life: Battery life is, undoubtedly, vital if the portability is to be of any value, and it, in turn, determines the period a radio set can be coped with between a charge or a replacement of a battery. A radio with a longer battery life is more favorable in outdoor activities or disasters during which access to electricity may be limited.

Durability and Build Quality: Radio durability and build quality are critical factors affecting its service reliability and life span, especially in a stressful or rough environment. Waterproof and weatherproof radios made of solid building materials and tight sealings ensure your favorite devices can survive in realistic water, dirt, and impact because of their hardiness.

PERFORMANCE COMPARISON

Assessing the performance of Baofeng radio models can involve more than mere features and other aspects such as quality and durability. Accurate tests and field huge-ups demonstrate how different models react in different conditions. undefined

Signal Clarity: Signal clarity in our context can be understood as the quality and intelligibility of the audio signals that are being transmitted. Radios of advanced noise reduction with high-quality microphone and speaker elements often produce the clearest and most natural-sounding audio quality. This will improve communication efficiency, especially in hazardous conditions where misunderstanding highly influences accuracy.

Sensitivity and Selectivity: Sensitivity and selectivity play primordial parts in radio signal reception of the weak ones and occupied adjoining channels rejecting. The ones with stronger sensitivity are more apt to detect low-level signals, and the ones with high selectivity are able to screen out the annoyances, including the other stations' programs when the airway is crowded and noisy. These peculiarities are crucial for ensuring that a message is accurate and robust in communication ecosystems with high levels of electromagnetic interference.

Range and Coverage: The range and coverage of a radio can be highly affected by several factors, such as the frequency band employed, transmit power, antenna configuration, and environmental conditions. Antennas with greater control and improved antenna designs can provide range and coverage. This means they can be successfully applied to long-distance communications and situations where network coverage is large enough. However, it is essential to emphasize that the surroundings, including terrain, obstacles, and atmospheric conditions, influence range and coverage. This can be achieved through performing range tests in multiple environments that give a fair appraisal of the radio's capability when exposed to various environmental conditions.

SUITABILITY COMPARISON

Here's a breakdown of common scenarios and the corresponding Baofeng radio models that may be suitable:

Amateur Radio Operators: For the white-eye operator that seeks a very functional transceiver in an emergency, models like the Baofeng UV-5R and UV-82 will be excellent as they provide a good blend of performance, accuracy, and price. These handheld transceivers applaud the ease of use, wide coverage frequency range, and compatibility with different accessories.

Emergency Preparedness: During emergencies and disaster response, rugged and waterproof radios like the British-imported Baofeng BF-F8HP and micro-sized UV-9R Plus prosper because they are made of durable materials that can withstand harsh weather conditions. These rugged radios resist spewing, blowing dust, and impact, wherein, in an emergency, communication is crucial.

Mobile Installations: Dual-band mobile radio Baofeng UV-50X3 is a good choice for vehicle-mounted and fixed base station installations as it can provide more remarkable transmit power and signal range. These radios are built to provide continuous coverage on the move and in stationary use, making them ideal for protective services, transport measures, and commercial operations.

Outdoor Activities: Frequently located in the outdoor environment by trekkers or athletes, the Baofeng UV-5R and UV-82 are ideal communication devices due to their compact and lightweight construction. These devices, such as radios, can be carried in porches, have ease of use, and have features such as weather channels and emergency alerts that make them the perfect travel for outside activities.

1.3 TECHNOLOGICAL ADVANCES

In the ever-evolving landscape of amateur radio technology, Baofeng continues to lead the way with its commitment to innovation and technological advancement. This section explores the latest technological advancements incorporated into Baofeng radios, highlighting the cutting-edge capabilities that set them apart in the market.

One notable advancement is integrating digital signal processing (DSP) technology into Baofeng radios. By leveraging DSP algorithms, Baofeng radios can achieve higher performance and efficiency levels than traditional analog radios. This allows for clearer audio quality, improved signal reception, and enhanced noise reduction, particularly in challenging environments.

Another critical technological advancement is implementing software-defined radio (SDR) architecture in select Baofeng radio models. SDR technology enables greater flexibility and adaptability in radio design, allowing for real-time digital signal processing and modulation. This opens up many possibilities for advanced features such as digital voice modes, spectrum analysis, and software-defined networking.

Baofeng has also embraced advancements in battery technology to enhance the endurance and reliability of its radios. Baofeng radios are designed to deliver long-lasting performance even in demanding operating conditions, from high-capacity lithium-ion batteries to intelligent power management systems.

TECHNOLOGICAL INNOVATIONS IN BAOFENG RADIOS:

Baofeng Radios has repeatedly innovated to add modern technologies to their radios, boosting their reliability and multi-functionality. Here are some examples:

Frequency Hopping Spread Spectrum (FHSS): Using the FHSS technique by Baofeng provides privacy of communication and resistance to any interference as the radio equipment switches quickly during the transmission process.

Digital Signal Processing (DSP): DSP technology serves the purposes of enhancing audio quality and eliminating background noise to make the voice more distinguishable in untidy environments.

GPS Integration: Some Baofeng models may have GPS receivers built into them, allowing precise location tracking and beacon mode for emergency purposes.

Cross-Band Repeating: The latest generation presents the cross-band repeat function, which enables it to operate as a hub station, thereby expanding the potential range of communication.

As Baofeng continues to push the boundaries of innovation, users can expect to see further advancements in areas such as software-defined networking, artificial intelligence, and interoperability with emerging communication technologies. By staying at the forefront of technological innovation, Baofeng ensures that its radios remain a versatile and indispensable tool for amateur radio enthusiasts worldwide.

In the following chapters, we'll delve deeper into the features, capabilities, and applications of Baofeng radios, providing comprehensive guidance to help users make the most of these remarkable devices. Join us as we explore the fascinating world of Baofeng radios and unlock their full potential for communication, exploration, and discovery.

2. BEYOND THE BASICS - ADVANCED OPERATIONS

Amateur radio provides an excellent opportunity for advanced operations to be more than just essential communication. This is achieved through advanced methods and technologies that assist in boosting communication quality, guaranteeing security, and improving the application of innovations. This chapter delves into three crucial aspects of advanced operations: Programming Excellence, Cybersecurity for Radio Flooding, and Innovations in Radio Applications.

2.1 PROGRAMMING MASTERY

Programming Baofeng radios might be complicated, requiring specific settings like frequency, channels, and more to fit everyone's needs. Though you could achieve basic programming by using the radio's interface manually, for advanced programming, you will require software tools to give you that level of flexibility and efficiency.

Download and Install CHIRP: The commonly used problem in programming your Baofeng radio is the CHIRP (Comms Help In Radio Programming), the most popular software tool. CHIRP is open-source software that supports many hardware models, such as Baofeng. Its programmer offers an easy-to-use interface for the channels, frequencies, and settings and the import and export data to and from the radio.

To start with CHIRP software, visit the official CHIRP website to download the latest compatible version for your Windows, macOS, or Linux operating system, or SCAN DIRECTLY QR CODE. When the download is done, open the CHIRP program with installation instructions and complete the installation on your computer.

Open CHIRP and Select Your Radio Mode: After installing CHIRP, you must use a compatible programming cable to connect your Baofeng radio to your computer. Ensure the cable is securely connected to your computer's radio and USB port.

Download Existing Settings or Start a New Configuration:

Launch the software on your computer to begin programming your Baofeng radio with CHIRP.

Once CHIRP is open, navigate to the "Radio" menu and select "Download From Radio" or "Upload To Radio," depending on whether you want to retrieve existing settings from your radio or configure new settings.

Choose your Baofeng radio model from the list of supported devices from the dropdown menu.

Enter Frequencies, Channels, and Settings: If you're retrieving existing settings from your radio, CHIRP will prompt you to download them to your computer. Alternatively, you can create a blank template if starting a new configuration. With your radio model selected and the configuration loaded, you can manually enter frequencies, channels, and settings into CHIRP. You can input these details directly into CHIRP's interface or import them from a CSV file if you have previously saved configurations or frequency lists.

Organize channels: To enhance organization and ease of navigation, you can organize channels into groups within CHIRP. Assign names and descriptions to each channel or group to help you identify them more easily later.

Review programming changes: Before changing onward, you should have re-read all the changes to the tune made with CHIRP. Verifying frequencies, monitors, and settings must be done twice to minimize errors and have concise and consistent parameters. Before installing the configuration files, ensure you have made all the necessary modifications and are satisfied with the overall programming changes. After that, CHIRP will prepare and upload the configuration to the Baofeng radio. If you still need it, attach the radio to the computer with the programming cable. Then proceed to the "Upload To Radio" submenu, which is accessible from CHIRP's "Radio" menu. Watch the display's commands to shift the settings from one side to your Baofeng radio receiver.

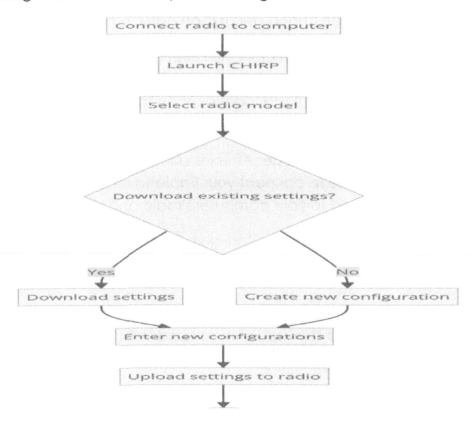

2.2 CYBERSECURITY FOR RADIO COMMUNICATIONS: ENCRYPTION AND SIGNAL PROTECTION

Cyber security is paramount in radio communications, for it highly guards information confidentiality and preserves communication paths' integrity. Though Baofeng radios are handy when it comes to communication effectiveness, they also introduce security risks that will have to be handled. This part concerns the theoretical concepts of encryption and signal protection inside the Baofeng walkie-talkies.

Encryption is the primary tool for shielding message transmissions from prying eyes. Encryption is an advanced means of encoding messages and ensures that only authorized people can decipher the messages, safeguarding them from being intercepted and unauthorized eavesdropping. When Baofeng signals lack basic encryption features, users can add external encryption devices or software solutions to improve their security.

One encryption software set with baofeng radios is Voice Encryption (VOX) software, which encrypts voice messages using sophisticated encryption algorithms. This software ensures end-to-end encryption of radio transmissions, meaning only authorized users who possess the decryption key can decode any encrypted data.

Signal security matters a lot when it comes to radio communications cybersecurity. Signal security techniques shall guarantee that data during radio transmissions will not be tampered with, eavesdropped, or interrupted by unauthorized parties. Baofeng radios may suffer interception of the signal and jamming, most notably in dense or hostile conditions.

Signal security can be improved through hopping frequency, spread spectrum modulation, and advanced error correction algorithms, creating a more secure environment. Frequency hopping simply means that during transmission, a swift shifting of frequencies occurs, which helps avoid being intercepted by unlawful listeners. Signal Spread Spectrum reconstructs signal among a wide frequency band, which leads to protection against the effects of jamming or interference. Error correction algorithms using sophisticated techniques have been built to correct the errors in the transmitted data so that communication can be established even in highly agitated atmospheric conditions.

Real-life applications of cyber security in foreign radio include professional work such as ambulance service, police work, and military operations. For instance, rescuers may talk amongst themselves using Baofeng radios to encrypt vital information when conducting deployment or disaster management operations. Public safety departments might use several signal security methods to protect the signal of their law enforcement during confidential investigations or surveillance functions. The military personnel may employ type-s-Baofeng radios that can be used for secure communication in the tactical mission or battlefield environment.

Generally speaking, cybersecurity is important since it ensures the security of essential elements such as confidentiality, integrity, and availability to work with Baofeng radios. Deploying encryption and signal security options would help users avoid listening in public spaces. Thus, unauthorized interception or disruption won't leak the users' information.

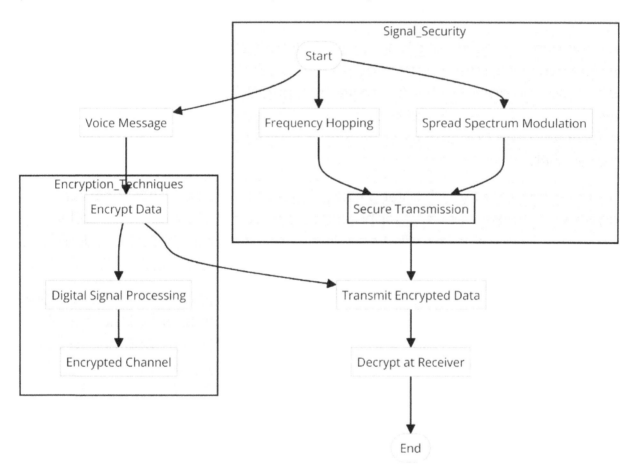

2.3 INNOVATIONS IN RADIO USE

From the public safety sector to the creative inventions that break the limitations, the versatility of Baofeng radios goes beyond traditional communication applications, enabling the system to grow and become part of various fields. The part focuses on how the handheld radios of Baofeng are generally used in up-to-date applications like drones, remote controls, and so on.

With the equipment of Baofeng radios, drones can enter a network of communication links with ground control stations, which in turn will result in real-time control and command of uncrewed aerial vehicles (UAVs). This possibility is also highly versatile and valuable in aerial photography, surveillance, search and rescue, and agricultural monitoring. Baofeng radios can bring a single connection, which includes a high perception, coordination, and responsiveness, into the drone system while addressing different operational situations. Emergency communicators, with the advent of the Baofeng radios, users can discover the new edges of the world as well as other fields beyond communications.

Regarding remote controls, Baofeng radios are an excellent example of a transforming feature. With remote vehicles, machinery, and operations, the Baofeng radios are an essential addition for transmitting control signals over long distances with higher reliability and optimized performance. This factor is of tremendous importance in industries such as agriculture, construction, and transportation, where such a mode of operation is standard. By offering a reliable communication channel for remote control systems, Baofeng devices break through the communication link to control and monitor in harsh conditions.

In addition, these radios are also used to save response, prevent artificial disorder, and protect public security. Moreover, their portability, usability, and affordability are key features that provide responders, respondents, and community-based emergency response teams with reliable communication. Baofeng radios can ensure real-time communication and grouping among emergency personnel; therefore, the actions of response and mitigation can be facilitated effectively in crisis situations. To better communicate the use of Baofeng radios, using example case studies from real work experience can provide a deeper understanding of the practical use and benefits of such radios. For instance, emergency medical services (EMS) agencies could employ Baofeng communications radios for real-time relaying of critical information during an emergency to improve patient care and timeliness. Moreover, fire departments would find Baofeng radios helpful for communication and command during firefighting incidents, thus optimizing situational awareness and the effectiveness of the operations.

It is evident that Baofeng radios have various creative uses in drones, remotes, emergency management, etc.; they are diverse and adjustable in complicated environments. Through the Efficient use of Baofeng radios by users passionate about trendy applications like communication, collaboration, and productivity, there is no doubt that radio technology will experience revolutionary breakthroughs.

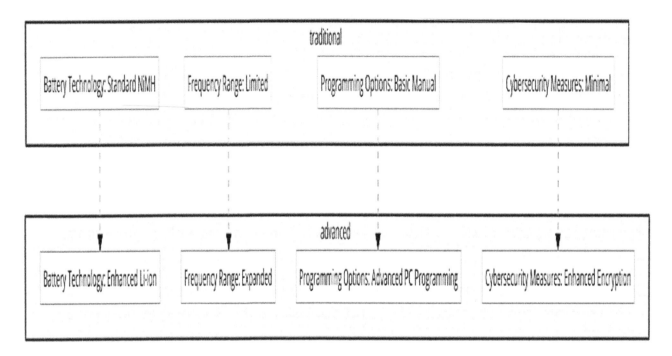

3. EXPLORING ACCESSORIES AND ADD-ONS

Baofeng radios offer various accessories and add-ons that can significantly enhance their functionality, durability, and overall convenience. This chapter will explore multiple accessories available for Baofeng radios and how they can elevate your communication experience.

Here is a brief overview of Key Baofeng Accessories:

Antennas

Frequency Range: Effortlessly transferable to VHF (136-174 MHz) and UHF (400-520 MHz) bands of operation. Antennae that are able to function with these frequency ranges are suitable for Baofeng IoT radios.

Connector Type: Most lit Baofeng radios come equipped with either an SMA or a BNC connector for antenna attachment. Verify that the antenna terminal is of the same type as your radio´s connector.

Antenna Gain: Decibel (dB) is the measurement unit for antenna gain and shows the degree of polarization. An aspect that higher gain antennae have to offer is the higher signal performance itself, but it may also be larger in size.

Impedance: 50 ohms usually forms the impedance value of the Baofeng radio antennas. Check to see that your antenna can stomp your radio's impedance to gain the best signal transmission.

Batteries and Chargers

Battery Capacity: Horizon's radio batteries are made in different dimensions with a battery capacity that is normally specified in milliampere-hours (mAh). With bigger capacitated batteries, you can get more hours of operation when the battery is fully charged.

Voltage: The Baofengs run largely on 7.4 V. Make sure the replacement battery voltage is either the same or lower than your radio's voltage requirement to avoid damage.

Compatible Charging Voltage: 8.4V is the normal charging voltage for the cellular battery of Baofeng radios. Employ chargers that give ample charging voltage to avoid overcharging problems and provide the right voltage for your device.

Charging Connector Type: They charge through USB or barrel plug connectors. Choose a charger with a connection type appropriate for the charging port used by your radio.

Charging Time: The charging time varies according to the battery power and type of charger being used. Considers manufacturer's guidelines for the exact accurate charging time that corresponds with the best battle life and the long life for the battery.

Cases and Holsters

Dimensions: You will find protective cases in different sizes made for Baofeng radio models. Examine the size of the case or holster to make sure there is no mismatch with your radio model.

Compatibility: Manufacturers, in fact, create cases and holsters exclusively designed for Baofeng radios. This guarantees a proper fit and provides quick access to the radio's controls and ports, which are both standard in this model. Try to find those match or bump cases and holsters that are compatible with your Baofeng radio model.

Features: Think of features like the clip attachments to the belt, the straps to carry, and the material durability while looking for a safety case or holster. Check available options that are more suitable for the purpose of using your radio and your environmental conditions to ensure the ultimate security of your radio.

3.1 ANTENNAS

In radio communication systems, antennas are essential to the transmitter-receiver process; they help improve the received signal quality and range and ensure the system's overall performance. The antenna is constructed to pick up radio waves and convert them from signals to waves which will be either sent or received. Baofeng offers assistance with most antennas so that one can choose what he favors according to his technological needs. Likewise, the longer antenna may work to expand the radio signal remotely, whereas a shorter antenna can reduce the radio's size and weight. Additionally, directional antennas might streamline the straight in only one direction, while omnidirectional antennas might send the signal in every direction. Baofeng radio users can optimize their radio communication system by selecting a suitable antenna for better performance and clarity. Here are the different types of antenna you might need: By choosing the appropriate antenna, Baofeng radio users can optimize their radio communication system for better performance and clarity. Here are the different types of antennae you might need:

High-Gain Antennas: An antenna with high-gain features has been mainly made to strengthen the signal and make communication more effective over a long distance. The antenna design applies an advanced engineering method to obtain a sophisticated signal capture zone followed by an improvement in the antenna's sensitivity towards a specific signal source. Thus, antennae with such sensitivity are an irreplaceable attribute of any radio device that may fail in the case of devices in remote areas or low signal areas.

High-gain antennas are crucial for long-distance communication as they can overcome the window or loss of signal naturally occurring over long distances. In addition, they have a lot of flexibility in that they can be used in conditions where connecting motorists is essential, for example, during communication emergencies and for stabilizing connections in critical communication networks.

In addition to their functional benefits, high-gain antennas are relatively easy to install and operate, making them a popular choice among users who require reliable communication over long distances. Overall, these antennas are essential for anyone who needs to communicate in remote or difficult-to-reach areas, and their importance cannot be overstated.

Dual-Band Antennas: Multiband antennae such as dual-band antennae are one of the most widely used antennae that work on multiple frequencies without any manual switchover. It covers VHF(Very High Frequency)and UHF(Ultra High Frequency). The ability of these antennas to work with different sources of communication is a sign of their versatility and adaptability in multiple situations. They are the workhorses of such applications, ranging from broadcasting television and radio, flying, military and emergency services, personal amateur radio, etc.

Dual-band antennas have a graceful way of working in various frequency ranges, and in cases where networks use two or more frequency bands, this transitioning of frequencies is essential. Such communication systems like radio, which operates at VHF frequencies, may not be able to communicate with another one that works at the UHF frequency. On the other hand, with a dual-band antenna, communication problems such as those can be bypassed entirely. This kills two birds with one stone ability, so users don't manually have to change antennas since this can be time-consuming and obvious sometimes.

As a tried and true measure, dual-band antennas solve connection problems for those operating in different frequency bands who must keep communications uninterrupted. They are the most advanced devices in the market today, making them an essential part of the pro and hobbyist community.

Tactical Antennas: The tactical antenna is communications equipment made to be weather resistant against the harsh climatic conditions found outdoors, making it suitable for use in such environments. These antennas are manufactured to handle weather and rough ground abuse successfully, so they continue to provide stable operation in harsh environments. These are durable and, for coating, have high-caliber designs, giving them a longstanding and reliable service. Tactics antennas are essential in camping and hiking equipment, are well suited for emergency call operations, and are perfect for troops requiring secure communication channels. Be it camping, hiking, or just an ordinary military operation you are engaged in, with tactical antennas, you have a suitable means of communication that can stand through any climatic condition you are in the middle of. Being of unyielding design and weather resistant, these antennas guarantee access with no limitations as long as you roam.

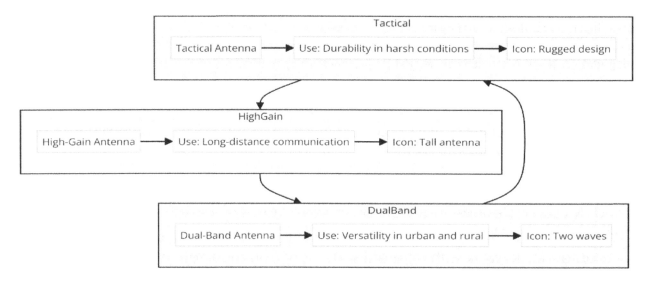

3.2 BATTERIES AND CHARGERS

It can be very frustrating when your Baofeng radios turn on. Thus, keeping extra batteries and chargers with you is essential to connecting with your Baofeng radio. As for the batteries of your radio, you do not know when the right moment to operate it may be; thus, you need to ensure it has been appropriately charged; otherwise, it may run out of crucial communication. So, the issue of spare batteries and chargers capable of quickly and safely charging up the Baofeng radio's batteries is always critical. By doing so, you can rest assured that you'll be able to communicate effectively without any interruptions or hiccups. Here are some essential accessories in this category: By doing so, you can rest assured that you'll be able to communicate effectively without any interruptions or hiccups. Here are some essential accessories in this category:

High-Capacity Batteries: Batteries with high capacity assure users that their radios will operate for a long time without recharging, which means they are vital for anyone who needs their radio to work for extended periods. This means that be it the case of an outdoor activity engaging a user, emergency response missions, or extra-long missions requiring a high-rated battery, safety and comfort are guaranteed since users remain connected 24/7.

Solar Chargers: Solar chargers are electrical equipment that can convert sunlight into electricity using photovoltaic cells and use this charge to recharge your Baofeng radio batteries. These chargers are the ideal copper-clad solution for any battery-powered radio, as they don't need an external electricity supply; eighthly, no pollutants are emitted. Harnessing the power of the sun can be a feasible solution for avid outdooring, outdoor activities enthusiasts, and emergency kits users as they have an option of using solar chargers, which are reliable and convenient devices to charge your radio even in an off-the-grid adventure, excursion, or emergencies where access to the grid may be less or absent. Moreover, the chargers are manufactured in different sizes and patterns; therefore, you can shop around no matter the type you need or want.

Desktop Chargers: Desk chargers allow us to charge our Baofeng radios and pair batteries in one go. We have created a battery charger with several bays for charging spares, so the operators can simply carry them along and recharge the used radios without downtime in communication. Desk chargers are the post-accompanies for offices and organizations that work with Baofeng radio to assist in daily operations or emergencies.

3.3 CASES AND HOLSTERS

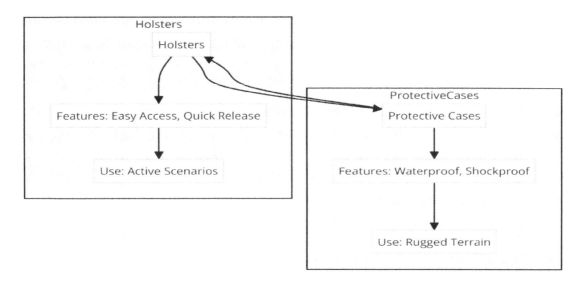

Putting your Baofeng radio guard against damage and wear is necessary to ensure its well-being, especially in outdoor environments and fieldwork. Here are some indispensable accessories in this category: Here are some essential accessories in this category:

Protective Cases: Protective cases are essential for Bakofeng radios to protect against drops, scratches, and moisture. These cases are designed to be resilient to the most demanding environments, implying their longevity and durability with your radio broadcast machine. They are constructed from different materials in various variants that meet specific needs and personal preferences. Such cases are engineered to provide unbeatable protection against damage while still retaining a high level of accessibility and ease of usability, enabling you to use the radio with utmost convenience.

Protective cases offer peace of mind and security against the rigors of daily use, whether you're trekking through the wilderness or navigating urban landscapes. Investing in a protective case can safeguard your radio investment and extend its lifespan, ensuring it remains in top condition for many years.

Belt Clips and Holsters: When it comes to having Baofeng with you all the time, I bet you know its vitality, protection, and ease of carrying wherever you go. **Belt Clips and Holsters** are accessories that help you look good in this outfit. It provides a secure bottle-ring design for holding your radio at the person or gear. As a result, you can keep your radio within reach at any time.

The belt clips are lightweight and tiny and can clip easily on a belt or backpack strap. They are very tight and protect your radio from misplacement, such as during moving or exercise. Conversely, the holsters are massive and provide your radio handover more safety. Most have a belt loop or clip connected to your waistband, backpack, or chest rig for easy carrying.

Attaching a belt clip-on or a holster lets you freely choose where to keep the radio and have quick access to it whenever needed. This is immensely helpful when hiking, camping, or participating in fieldwork, and fast communication is often the most crucial thing. With a clip or holster, you can always keep your walkie-talkie handy and assume your radio is always with you to react promptly when needed.

3.4 HEADSETS AND MICROPHONES

Headsets and microphones are a must- accompanying equipment to help you elevate communication clarity, privacy, and usefulness. Here are some noteworthy accessories in this category:

Surveillance earpiece: Surveillance **earpieces** are a must-have tool for security personnel and those who work in law enforcement. It also is of the operators who work undercover. Through these devices, individuals communicate discretely without simultaneously occupying both hands, allowing them to stay on top of their game physically and mentally. The very highest-quality earphones are made to block out or nullify background noise and unwanted distractions; thus, end users can communicate efficiently in essential and difficult environments. The earpieces can vary in shape and size and can be customized to maximize cost and usefulness. This is because they are equipped with top-notch unidirectional microphones and high-performing speakers that capture precise and reliable voice transmission, even with a lot of background noise. The earpieces are designed to meet an even more covert goal, being of tiny and obscure form factor that largely remains invisible to the naked eye. Similarly, an earpiece with its advanced functionality and potential is a prime choice for those who must keep in touch smoothly and secretly if the circumstances demand it.

Speaker Microphones: Speaker microphones are a class of earphones capable of producing high-quality sound and simultaneously, can receive input from outside the ear. These devices create a smooth pathway for speech to be carried out uninterruptedly in noisy or crowded situations, including clubs and security activities. The inbuilt speakers create an elevated volume and a vivid presentation by boosting the sound level and clarity, whereas microphones engineered to block out background noise ensure that communication is clear and steady without interruption. These devices are of great value to law enforcement, security personnel, event organizers, and advocates for outdoor activities who, in their duties, often be in difficult and unsafe situations where problems can quickly occur, and communication becomes an essential tool.

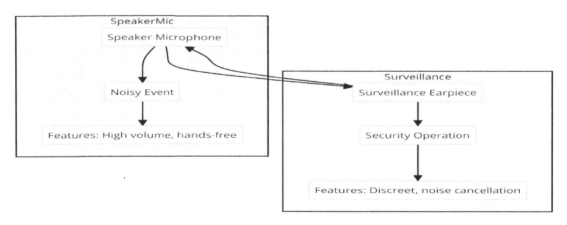

3.5 ACCESSORIES MAINTENANCE

Choosing the proper accessories not only plays a key role but the users must also be trained on how to maintain them to preserve their durability and full capacity. Here are some maintenance tips for different accessory categories:

Antennas

Regular Inspection: From time to time, inspecting a vent for any signs of damage, like bent elements, cracks, or rust on connectors, is also important.

Cleaning: Clean the antennas with a soft cloth loaded with mildly soaped water frequently. Wash the antenna lightly with a lint-free cloth to remove dust, dirt, and debris that may affect signal transmission by the accumulation of these.

Avoid Harsh Chemicals: Opt for cleaning products that contain mild chemicals and gentle abrasives instead of harsh chemicals or abrasive cleansers, which can wear down the coating or the connectors of the antenna. Keep using softer solutions and gentle wiping to reduce collateral damage.

Proper Storage: During the storage time, place the antennas in the instruction case or sleeve to prevent them from getting damaged and ensure they stay away from dust and rain. Keep the antenna cable straight and not bent when not in use.

Batteries and Chargers

Avoid Overcharging: Wear batteries will have a shorter battery lifespan and lower overall capacity when you always leave them on the charger even after they are fully charged, so remember to take them off once they are fully charged.

Proper Storage: Store the batteries in a room that's not sunny, heat, and humid. To avoid doing so, avoid storing batteries in humid or moist areas, as exposure can lead to corrosion and potentially long-term damage to the batteries.

Cleaning Contacts: Check and clean up the contact charging on both batteries and set up a good electrical connection. With a soft, dry cloth, remove any dust and debris that may stain the contacts caused by excessive residue.

Use Genuine Accessories: Apply only genuine Baofeng batteries plus chargers with compatible accessories for third parties or approved by the manufacturer. Legally authorized and original production may damage radios and cancel the warranty coverage for the same.

Cases and Holsters

Regular Cleaning: Clean and wipe the covers and holsters with a damp cloth or sponge dipped in mild soap and water solution at least once a week. Here, you can use your eyes to notice any stains and debris that accumulate during the product's use and then wipe the surface gently.

Inspection for Wear

It is recommended to inspect cases and holsters periodically for signs of wearing conditions, for example, a ripped seam, torn product, or disengaged clips. Make sure to dispose of damaged or worn-out cases and replace the radio with new cases to maintain efficiency and protection.

Proper Fit

Make sure that the radio fits tightly into the baggage or holster so that there is not too much movement or rattling. Providing a good inside fit of the radio not only proffers the benefit of likely decreasing damage to the radio but also allows for better protection during accompaniment or use.

Besides the radio models they depend on, Baofeng users can explore and invest in accessories and add-ons to tailor their communication setups to suit their particular requirements. Be it outdoor activities, emergency preparedness, or professional use; these accessories are developed to increase the functionality, robustness, and convenience of Baofeng radios, which make them suitable for communication in any situation.

3.6 USER TESTIMONIALS

"Lately, my wilderness expeditions have become thrilling because I now have a Baofeng radio with a high-capacity battery. The advantage of longer operating time has been one of the greatest; when I am off the grid, I no longer depend on radio power. This is a reliable partner that connects me with my crewmates and ensures my safety during our adventures." - Sarah W.

"One of my best purchases for outdoor activities has been a protective case for my Baofeng radio. Whether I'm hiking or camping, my radio case saves it from getting beaten around, dropped, or wet, and it's sturdy enough to carry while biking. In addition to being rugged, it's also lightweight, and you won't need a special bag to carry it along with other essentials.

"While serving as a volunteer emergency responder, the solar charger for Baofeng radio has been a real savior. You may not be able to rely on traditional power sources during extensive search-and-rescue missions. It's a grand warranty to have this reliable solution to deliver first-rate communication times when seconds count."

"I cannot even fathom operating my BAOFENG radio without my desktop charger. The multi-battery charging facility is an added benefit where it is quite feasible to load a couple of batteries at the same time. Used at the time of duty or on leave, this desktop charger will always be at my disposal to keep my radio its much needed power."

These user testimonials have particular value due to the fact that they offer a first-person view as to how different accessories of the Baofeng radios have empowered communication experiences on several occasions. Through this, those advantages and benefits are given a splash of light as readers decide what to get for their own accessory investments.

3.7 CASE STUDIES

Search and Rescue Mission:

Scenario: Operating a rescue party in a location far from the usual power supply becomes a great challenge when the use of Baofeng radios for a protracted period depends on conventional power sources that are remote.

Solution: The team intended to use solar chargers as a means to harness solar energy, which would, in turn, be able to keep radios up and running for the whole mission.

Outcome: Thanks to the installation of these solar chargers, the search and rescue team successfully managed communication among the members of the team and base stations, thus facilitating timely response and the most efficient coordination during critical moments. The introduction of solar chargers to our team made it a reliable and sustainable energy source. Thus, everybody could focus most of their time on their mission rather than worrying about the battery life.

Emergency Preparedness Drill:

Scenario: The emergency preparedness regulation held a drill to rehearse the possible scenarios amidst a natural disaster as the response strategy. A specific challenge was to make sure that the staff was always connected virtually from the consigned environment.

Solution: The organization designed the Baofeng laptops with high-capacity batteries to extend operating hours, and the program stocked up chargers for the batteries for long periods.

Outcome: After the drill, you will not have any communication disruption as the whole team had the high-capacity batteries and desktop chargers with them. The long battery life resulting from high-capacity batteries gave assurance that the communication was uninterrupted during the training, while the desktop chargers eliminated any downtime that would be otherwise faced in between the tasks. This resulted in the organization successfully experimenting with commercial scenarios and the weak points of emergency communication protocols.

These case studies prove that Baofeng radio accessories, which involve solar chargers and capacious batteries, may work well in a diverse range of communication schemes to cover existing obstacles and reach end goals in situations of real-world problems.

4. IN-DEPTH BAOFENG CUSTOMIZATION

A Baofeng radio is highly admired for its unique characteristics, such as versatility, adaptability, adapting to different scenarios, and options for making changes and adding features. We will go to that chapter to see the miracles of how advanced Baofeng optimizes its radios for the best performance for you. In the next section, we will cover the range of antennas available on the market and offer guiding principles for choosing the right type for your requirements. Moreover, you will be instructed to handle power efficiently to ensure that you can maintain the best performance of the radio. At the same time, the battery is powered slowly. Ultimately, by the end of this chapter, you will acquire a complete skill set that will let you adjust the bands and channels how you want, making it a reliable, trusted device for your communication needs.

4.1 TAILORED CONFIGURATIONS: HOW TO CUSTOMIZE RADIOS FOR PERSONAL NEEDS

Personalizing the Baofeng channel involves reconfiguring a few settings and parameters to tailor your selection to personal preferences and needs. Be it a beginner at radio use or a pro, knowing how to set it up with its configuration is essential for using it to its potential and maximizing its performance.

Here are some critical aspects of tailored configurations for Baofeng radios:

Frequency Programming: Baofeng radios are multi-band receivers that cover the entire spectrum of VHF and UHF bands. Programming frequencies make it possible for users to get and communicate via channels. Regardless, suppose you are into amateur radio or emergency services or need frequencies for other purposes. In that case, you must program the audio-relevant frequencies according to your needs.

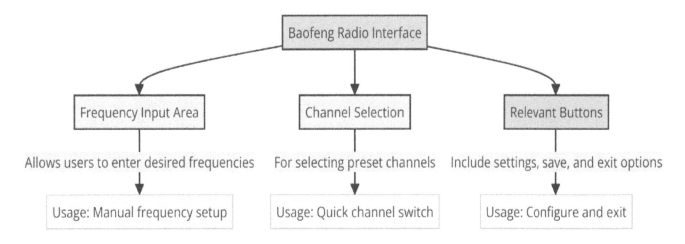

Channel Organization: Certain channels could be grouped by functionality or location for easier search and navigation, making it even more convenient. For example, this might be grouping for repeaters, channels, or frequencies that puts you near the transmission source. Assigning names and descriptions to channels gives additional organization and maintains simplicity of use.

Squelch and VOX Settings: A correct adjustment of squelch levels and VOX (voice-operated transmission) settings means a desired sound quality and transmission quality. The suppression technology allows the intercom to identify and eliminate the background noise without a signal. In contrast, the VOX mode automatically initiates the transmission when the operator speaks into the microphone.

Display and Keypad Settings: Personalizing displays and keypad settings leads to a user-friendly interface that suits the user in this manner. Balancing backlighting levels, contrast, and sound volume allows the illumination to suit different lighting situations.

Advanced Features: Baofeng radios have various advanced features, including dual watch mode, detector, and DCS/CTCSS (Digital-Coded Squelch System/Continuous Tone-Coded Squelch System) programming. These capabilities allow you to set up the functions of your radio according to your preferences and control their performance.

Individual fittings to users' preferences optimize their radio performance and operability. As a result, users can enjoy trouble-free communication under any circumstances.

4.2 ANTENNA SCIENCE: ANTENNA SELECTION AND TUNING FOR OPTIMAL PERFORMANCE

The antenna is an important element of a radio device for its successfully direct and undistorted transmission and reception of radio signals. Picking and setting the suitable antenna correctly is essential when conducting Baofeng radio performance at a sufficient distance.

Here's a detailed guide to antenna selection and tuning for Baofeng radios:

Antenna Types: Baofeng radios can come with free-hand antennas, which are small and light but might not give the strongest signal and furthest range qualities. Users can not only extend their radio's capabilities, but they can also do it by upgrading their antenna to a higher-gain antenna, like a whip antenna, or permanently installing the antenna to a pole or car mast.

Antenna Length and Frequency Regarding the operation of the Baofeng radio, the length of the radio antenna defines its performance most. For the rule of thumb, antenna length should be correlated to the resonant frequency increase. Baofeng radios are designed to be active within a designated frequency band. They carry an antenna with the best bandwidth to suit the radio frequencies and offer maximum performance.

Nevertheless, it must be admitted that the size of the antenna has a significant impact on its function in the lower waves. More giant antennae function best at lower frequencies, and shorter antennae are more practical at higher frequencies. Therefore, we conclude that choosing the appropriate length antenna for your aimed frequency band is essential to avoid performance losses. The longer the length of the antenna, the higher its resonance frequency. Baofeng radios grab some frequency band and, in this case, use an antenna to filter the bandwidth of the desired frequency to enhance the device's performance. The length of the antenna for an antenna is the primary factor that decides the performance on lower frequencies. On the other hand, when working on higher frequencies, a shorter antenna will be preferable.

Antenna Gain: Antenna gain is the degree of the angular spreading of an antenna radiation in a specific direction. With the added gain, larger gain antennas can help signal propagation and increase the range of Baofeng radios, mainly when used outdoors and in long hauls. However, what can be the compromise between benefit and portability above gain growth, as the antennas with greater features might be bigger and bulkier?

Antenna Tuning: Fine-tuning canopy length or configuration is everything for antenna aligning to the desired frequency. Dueling Baofeng radios usually work within a given range of frequencies. Therefore, appropriate antenna tuning provides better efficiency and more effective signal production. The antenna can be fed via an antenna analyzer or a SWR (Standing Wave Ratio) meter to measure and adjust for best tuning - impedance and resonance.

Antenna Placement: It also matters where the antennae are positioned since this strongly influences their functions. For handheld Baofeng radios, a good approach is to show the radio upright and stick the antenna vertically, strengthening the reception signal and transmitting more signals. Concerning mobile or base station installations, a contrasting antenna configuration by mounting at a higher position, for example, on a mast or a roof, will ensure better line-of-sight coverage and range.

With the proper antenna selection, tuning it well, and placing it at the best location, radio owners will experience much better performance and coverage of their Baofeng radios, which are very helpful for long-distance communication in different environments.

4.3 BATTERY AND POWER MANAGEMENT

Quality battery and power control are critical factors in maintaining smooth wireless communication and increasing the Baofeng radio's autonomy. Introducing power usage strategies that ensure optimal power consumption and intelligent battery development will improve radio performance and reduce downtime.

Here are some strategies for battery and power management with Baofeng radios:

Battery Selection: Like most other Baofeng radios, they have replaceable removable rechargeable lithium-ion batteries, which provide a long lifespan, are lightweight, and have power density. Customers would be able to go for the battery packs offered through the factory, or they could opt for ones with better quality for longer runtime. One should never go for cheap and unreliable batteries; instead, one should use reliable batteries from reputable manufacturers to guarantee quality and compatibility.

Battery Charging: Correct charging practice has a key role in the battery's durability over the years and in ensuring expected performance throughout its operational life. In most cases, the packages of Baofeng radios usually include a desktop charger or a standard USB cable for recharging the radio's battery. Avoid overcharging or undercharging the battery, as it will wear out the battery quickly and reduce its operational capacity. According to the manufacturer's recommendations, battery life will be prolonged regarding charging intervals and procedures. From:: Language Proficiency

Power Saving Features: The power-saving functionality of Baofeng radios allows for a considerable amount of work time even under regular, intense operation. Some enhanced capabilities may include regulated transmit power levels, sleep mode utility with low power consumption, and automatic power-off timers. Turning down the transmit power when the maximum range is not essential and encouraging power-efficient modes during non-activity time will minimize the user's battery consumption; hence, the devices will run longer.

Backup Power Sources: For the primary battery packs and alternative power sources for backup and extended use, users should consider single-cycle battery packs. External battery packs, solar chargers, or portable power banks can offer longer battery juice if you are in places where power is not always available (remote locations) or during an emergency (First aid). To be on the safe side, choose backup power unlocking devices with the needed overcapacity as well as radio compatibility.

Battery Maintenance: Proper battery maintenance is the most critical thing. It isn't advisable to store and operate your batteries at extreme temperatures, in moisture, or any physically damaged state, as these factors affect battery health and its lifespan, resulting in the substandard performance of your device. Inspect the battery from time to time, visually looking for signs of wear, damage, or any abnormalities, and, as necessary, replace the battery to ensure that the equipment works properly.

Therefore, complete Baofeng customization needs picking fluid configurations, antenna optimization, and incorporating excellent battery utilization and power management into these principles. When applying all such individualization techniques, users can extend the functional scope, duration, and accuracy of their Baofeng devices during speech transmission in different situations.

5. LEGAL INSIGHT AND GLOBAL COMPLIANCE

When talking about radio communication, we need to be familiar with legal regulations and universal compliance norms that determine the global use of communication devices. Like all other communication devices, these radios have numerous stipulations and limitations that should always be followed for the sake of the user and the security of the general public.

In this chapter, we will provide a thorough explanation for navigating the complicated areas of regulations that control the employment of Baofeng radio cups. It encompasses ethical aspects, responsible methods of using devices like Baofeng, and hints about global usage of Baofeng radios. Close attention to these tips and abiding by the legislation allows you to act on Baofeng radio even if you evade legal or compliance problems.

5.1 NAVIGATING REGULATIONS

The radio use and regulations rules can be complicated and differ tremendously from country to country. The use of Baofeng radios requires that the users understand and abide by the local code of practice when the radios are in use to avoid legal non-compliance and ownership of public responsibility. This affects all radio operations, including amateur transmission, commercial radio applications, and personal communication.

The purpose of radio communication in many states is to have a radio usage license, which can be received from the appropriate regulator. Depending on whether the user wants to conduct amateur operations, public safety, or business activities, these licenses may be split into subgroups. Users need to discover local policies and determine whether they have permits or authorizations from the rightful authorities before using Baofeng radios, which are illegal.

Regarding amateur radio operators, the mandatory step of getting a license is often what they need before legally using Baofeng radios. Amateur radio licenses represent the level of proficiency in radio operations and the knowledge about relevant regulations. These ensure that qualified individuals and responsible radio operators use amateur frequencies.

Besides licensing criteria, users are also forced to abide by the frequency allocation rule and the measures adopted by national regulatory agencies. Baofeng radios are tuned to many frequencies, including VHF and UHF bands, with each band allocated for amateur radio use, public safety services, or commercial channels. It is very vital that the users are authorized frequency band users and comply with the frequency coordination guidelines, as a failure to do this may result in interference and, consequently, legal implications.

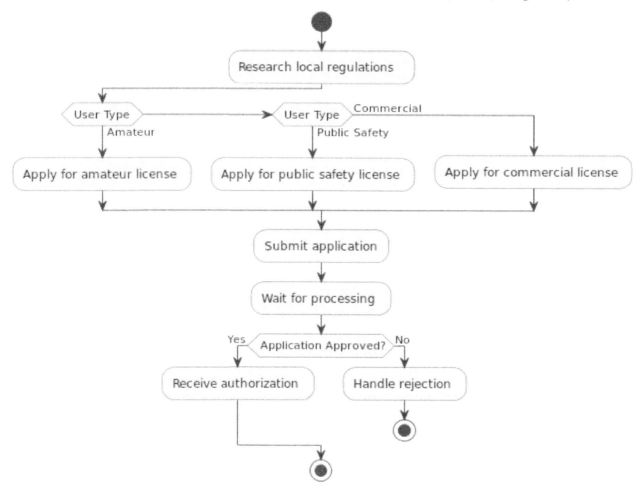

Also, the users should know that there are limitations on transmit power levels, modulation types, and encryption capabilities governed by regulatory agencies. These constraints are established to maintain fair application of radio frequencies and mitigate interference between users and services. Users must follow these restrictions to prevent legal consequences and, on the other hand, to maintain the safety and security of other radio users as well.

It not only entails fulfilling legal obligations but also a moral obligation. Adherence to the licensing obligations, frequency allotments, and operation regulations leads to good spectrum utilization, and a sense of responsible communication conduct culture shall be nurtured. It is essential that users continually educate themselves on local rules and make sure that they use radios responsibly and as prescribed by the law.

5.2 ETHICAL PRACTICES: ADVOCATING RESPONSIBLE COMMUNICATION ETHICS

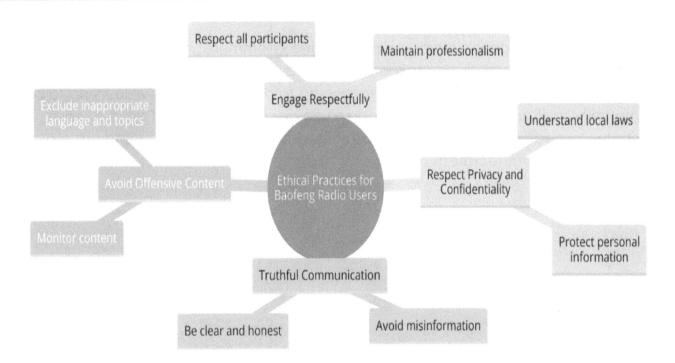

Ethical communication practices are essential in creating an atmosphere of culture, integrity, respect, and professionalism in the media industry. In today's technology, where everything moves quickly, Baofeng radio users must be responsible enough to use ethics and be cautious when dealing with technology to avoid misuse or abuse.

The fundamental ethical rule of radio communication is to treat the privacy and confidentiality of communications properly. The abuse of the system, namely unwarranted listening or interception of private conversations without consent, is the most serious issue because it infringes the privacy rights of individuals and destroys trust in the communication system. It is of extreme significance if essential information is under seal.

Another ethical aspect is to stay truthful and sincere in communications. Users must truthfully declare their details and intentions when transmitting at radio frequencies to avoid the risks of misrepresentation, which could otherwise cause misconception or wrongful interpretation. The principle is significant in settings where prompt and clear communication is fundamental and avoids misinterpretations.

Furthermore, users need to be on guard not only when revealing confidential or private information over the radio but also when checking the content of their communication to make it appropriate for the outlined target audience and circumstances. The transmission of offensive or abusive content and any form of prohibited content should be avoided to avoid transmitting a professional and respectful communication environment. People are advised to exercise vigilance because their choices critically affect the environment and the radio spectrum. The rules governing frequency synchronization, not interfering with other users of the frequencies, and compliance with all the required regulations reflect rightful and ethical radio communication.

Responsible communication ethics, along with one's communication with other radio users or operators, should also be taken into consideration. Those who use the radio should be polite, cooperative, and understanding to other operators while trying to build a community spirit and relationships among radio operating people. This, in a sense, contributes to peaceful surroundings, which allows everyone to enjoy being among others and feeling appreciated.

5.3 INTERNATIONAL USE: TIPS FOR TRAVELERS USING BAOFENG ACROSS BORDERS

For travelers who carry Baofeng devices during trans-border trips, it is necessary to understand the relevant regulations/procedures and comply with them if they want to operate legally and hassle-free. undefined

Research Local Regulations: Before traveling to a new country, spend some time on the internet to find out what the local radio usage laws and regulations about using Baofeng radios will be. While this authority is in charge of licensing, frequency allotment, and imposed operational restrictions, pay attention to this.

Obtain Necessary Licenses or Authorizations: When essential, get the required licenses or approvals from the legal authority in the destination country to run the Baofeng radios legally. Non-commercial radio operators need to ensure that their stations' rights are recognized or are part of the reciprocal agreements when operating internationally.

Check Frequency Compatibility: Ensure your beloved Baofeng radio operates on the frequency range available in the destination country. Some countries employ different frequency ranges or use different modulation types. This means that radio users will have to change their settings accordingly.

Respect Local Customs and Practices: Observe a place's norms, traditions, and cultural variations when using the Baofeng radio in a foreign country. Do not forget the language barrier, communication protocols, and cultural aspects to prevent conflicts or miscommunication between locals and authorities.

Exercise Caution in Sensitive Areas: When using Baofeng radios, operate them carefully so that they won't cause any security breaches in sensitive areas such as military installations, government facilities, or border regions—steering clear of dumping stations near critical spots or participating in any suspicious activities that might attract the attention of local authorities.

Carry Documentation: Be sure you have copies of the related documentation, such as the radio license, frequencies allocated to you, and identity documents, while traveling with the Baofeng radios. The visibility of these documents can be implemented on request to show the business has considered compliance with the relevant local regulations and in case of further communication with the authority.

Plan for Emergency Communication: Further to constant communication needs, the extras of emergency communication capable of radio traveling with Baofeng must be noted. If necessary, reprogram your radios to be tuned to the emergency frequencies, channels, or contacts involving local emergency services or support groups.

Following the above recommendations and regulations, travelers can operate a Baofeng radio safely and correctly without a license in many countries. They can communicate freely and legally within the borders of the nations so that they meet the local laws and requirements.

5.4 CASE STUDIES AND EXAMPLES:

To provide a clearer picture of the consequences of legal compliance when using Baofeng radios, let us look at two real-life disadvantages that entities or individuals have experienced.

To shed light on one particular case, the resident of a highly populated urban area who had been a hobbyist in the use of a Bafeng radio was fined and faced legal action from relevant authorities as he never obeyed the relevant license rules. With their unauthorized broadcasting, they managed to shut down the important communication channels for emergency services, hampering the communication between emergency personnel and risking putting the public in danger again. This case indeed is an eye-catching reminder of the necessity of learning the necessary legal rules and following them to avoid legal consequences and road use responsible means.

An example, too, is the individuals who incorporated the Baofeng radios into their commercial business without having the licenses. In the face of protests by regulatory bodies, the organization nevertheless did so thoughtlessly, integrating radios for internal communication and coordination without the respective authorizations. This outcome entailed huge penalties that ranged from court fines to temporary shutdowns of its operations. This depiction reminds us of the legal consequences of operating a Baofeng radio in business operations and the possible outcomes of disobedience to the law

5.6 OBTAINING LICENSES

To simplify the process of obtaining licenses for Baofeng radios in communities, these are tips for the US and other countries. One of the first things to cover will be the information about the way the license is issued in the United States. Then, we will touch on the topic of criteria for issuing licenses in other countries as well.

United States:

In the United States, people can get their mdash; Baofeng radios licenses from the Federal Communications Commission (FCC). The FCC controls the allocation of radio frequencies by licensing specific radio communications to individuals and organizations.

Here are the steps involved in applying for an FCC license:

Determine the Type of License: The FCC issues different types of licenses based on the purpose/usage intended for the radio equipment. Usually, licenses fall under different classifications, such as the General Mobile Radio Service (GMRS), the Amateur Radio Service (Ham), and the Commercial Radio Operator License (for marine and aircraft use).

Fill Out the Application Form: Whether the license type is an amateur radio operator license or broadcast license, the applicants must fill out the application form provided by the FCC appropriately. According to this, most of the forms comprise personal details, data regarding radio gadgets, and fee payments.

Pay Fees: On the FCC licenses, some of the application fees are necessary. The fees vary from a single license to a license of validity and duration. The applicants can submit their fees for the application online using the FCC's Electronic Comment Filing System (ECSS) via the web or by post.

Pass Examinations (if applicable): The amateur radio service license, for instance, may necessitate an examination as one of the requirements, mostly aimed at demonstrating technical competency and comprehension of regulations. Some recognized organizations provide study manuals and examinations for Amateur Radio operators.

Submit the Application: Based on the form that has been complied with and any fees paid, the applicants can finally submit their filled application electronically through the ULS or by mail.

Await Approval: The FCC is to review the application and then decide if the license will be digitally transmitted or sent in the physical form of a letter. Licenseholders shall follow the licensee's terms and conditions and the FCC directives regarding radio operation.

You can reach them on the FCC's official website or call them directly if you need more information about the rules associated with the FCC license application.

International:

Whereas for people dwelling overseas, the license acquisition process could be different according to the country's system regarding its regulations. As a rule, research must be conducted to acquire relevant knowledge of the local licensing conditions and procedures. It must be considered that these may differ across the country in each jurisdiction.

In essence, the user can contact the national communications regulatory authority for telecommunications or radio communications to get information about the licensing procedure. Plans and procedures that control workforce movements and regulate the cordoning area are typically required.

5.7 PROFESSIONAL INSIGHTS

To enhance the discussion on legal insights and global compliance, we will integrate insights from legal experts and seasoned radio operators specializing in regulatory compliance:

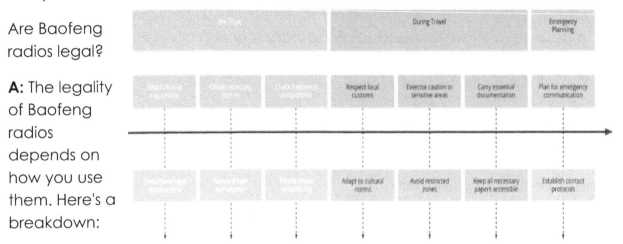

Are Baofeng radios legal?

A: The legality of Baofeng radios depends on how you use them. Here's a breakdown:

For Amateur Radio Operators: Baofeng radios are legal for use on licensed amateur radio bands (with a valid Amateur Radio License).

For Other Frequencies: Using a Baofeng on frequencies outside amateur radio bands (e.g., FRS, GMRS) is illegal. These radios aren't certified for those bands.

Q: I have a Baofeng. Do I need a license?

A: You only need a license if you're using the Baofeng on amateur radio frequencies. In most countries, a government agency regulates amateur radio licensing. In the United States, it's the Federal Communications Commission (FCC).

Q: Can I modify my Baofeng to use other frequencies?

A: Modifying a Baofeng to use unauthorized frequencies is illegal. Tampering with the radio can cause interference with other transmissions.

Q: What happens if I'm caught using a Baofeng illegally?

A: Penalties for illegal radio use can vary depending on the specific violation and your location. They can range from fines to confiscation of equipment.

Q: Where can I learn more about amateur radio licensing?

A: Your local amateur radio club is a great resource. You can also find information on the FCC website https://www.fcc.gov/ or the website of your country's communications regulatory agency.

6. REAL-WORLD APPLICATIONS AND CASE STUDIES

While the primary use of Baofeng radios is a simple communication medium, there are more things they could be used for because of their functionality and capabilities that are inclined towards increasing the efficiency and effectiveness of both individuals and organizations. It does not matter whether it is during an emergency preparedness, professional setting, or community relation job, Baofeng radios are highly versatile and ready to take the challenge when the need arises. This chapter demonstrates comprehensively how Baofeng radios are being used effectively with other applications and case studies that prove that Baofeng radios are the best practical and advantageous devices in multiple situations. The practical demonstration of the above examples by people or organizations can be an eye-opener in exploiting the full potential of Baofeng radios and attaining the targets and objectives.

6.1 EMERGENCY PREPAREDNESS

6.1.1 CREATION OF STRONG COMMUNICATION STRATEGIES

For places highly susceptible to disasters such as hurricanes, earthquakes, and forest fires, a disaster communication plan is necessary to quicken the emergency dispatching measures that need to be followed. The devastating consequences of the 2005 Hurricane Katrina should be presented as the foremost reason for attention here. The communication among diverse businesses was very effective, and in the end, it had the upper hand. During the Baofeng Hurricane, two-way radios were employed constantly by everyone from the emergency response and the volunteer sector up to the affected communities. The wireless devices allowed the command staff to make contact with anyone they needed to and transported food, medical supplies, and rescuers to the survivors.

Respondents must develop flawless communication norms, identify primary and alternative channels, and prepare communication equipment before a disaster strikes. This horizon includes:

Regularly carrying out device maintenance and communications testing.

Having backup power generators.

Teach all crew members how to use the tools correctly.

The following ways of acting can help responders keep lines of communication open and working, even during the most robust turbulence.

Providing adequate communication strategies and tools is critical to adequately managing the impact of natural disasters. Being Clear and Reliable with Communication helps responders coordinate their work to support those who require assistance more effectively.

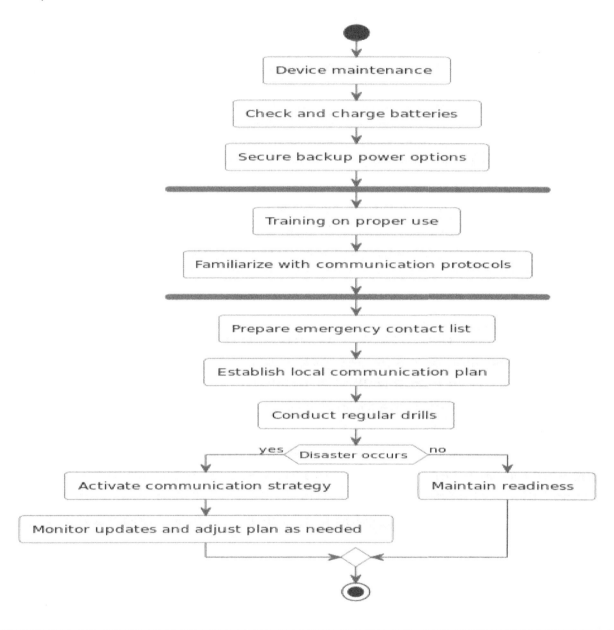

6.1.2. TESTIMONIALS AND REAL-LIFE EXAMPLES

At the end of the horrible earthquake in Nepal in 2015, Baofeng radios revealed themselves as irreplaceable instruments of assistance in mobilizing relief work. Testimonies of rescuers and medics emphasized Baofeng radio's advantages in creating communication networks in recesses where natural access is killed. Such radios allowed for the instant communication of rescue squads, medical centers, and logistics among themselves. This significantly facilitated the well-coordinated use of resources and provided timely supplies to the targeted people.

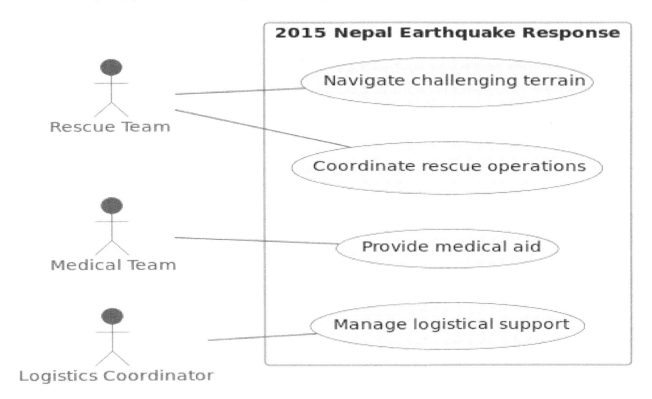

Thanks to their long transmission range and high capacity, the Baofeng radios could similarly surmount the challenges of the steep terrain and the poor infrastructure that had hindered other radio providers. The radios could always be assured of reliable communication channels amidst the isolated areas in which rescue units were operating. This helped them coordinate their efforts better, as their operations were more effective.

Also, the Baofeng radios contributed significantly to ensuring that affected people got medical care. The medical workers utilized radios to communicate with other medical personnel and logistics centers in such a way that they were able to obtain the required supplies, such as medicines, bandages, and stretchers. This simplified the process of delivering medical assistance to the people who needed it the most and ensured that people got it.

The use of the Baofeng radio receivers during the 2015 Nepal earthquake was a testament to their reliability, functionality, and adaptability in creating communication systems in stressful surroundings. Their work in disaster relief efforts has demonstrated their remarkable effectiveness in saving countless lives. They also proved their crucial role in modern disaster response and relief efforts.

6.2 BAOFENG IN PROFESSIONAL SETTINGS

6.2.1. SECURITY AND EVENT MANAGEMENT

The Olympic Games' scale, one of the most significant sporting events globally, needs substantial safety and security plans to guarantee the athlete's and fans' safety. Of course, Baofeng radios played a crucial role in such a huge event and enabled successful operations management.

Before the last Olympics, the Baofeng radios were widely used by safety staff, the organizing committee, and the medical team. These radios allow the teams to communicate effortlessly and synchronize their activities, thus aiding the coordination and resolution of emergencies.

Regarding crowd management, Baofeng radio is a powerful tool that monitors security threats and helps solve any possible problems immediately. They became a trustworthy and fast communication source that helped security operatives take action or start to do something before the incident happened.

The deployment of Baofeng radios forms a central part of the measures to keep the Olympic games safe, which guarantees the visitors a memorable experience during the event.

6.2.2 INDUSTRIAL AND COMMERCIAL APPLICATIONS

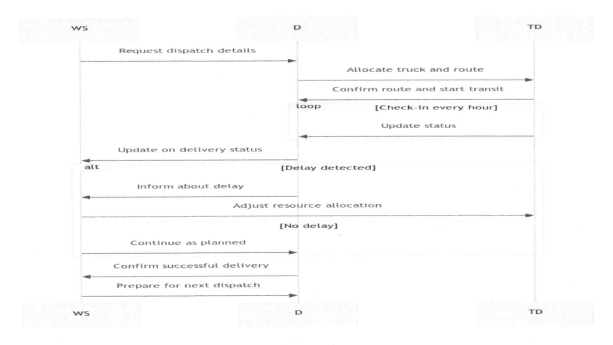

In logistics, communication is the backbone; there are constant communications surrounding warehouse, inventory, and delivery fleet management. A Baofeng walkie-talkie would be the most essential tool to help overcome this challenge. These radios are universal within logistics companies worldwide to guarantee smooth communication between warehouse staffers, truck drivers, and dispatchers whose goal is speeding up operations, minimizing delays, and optimizing the link between the resources and the utilization of the same.

For instance, one can think of a multinational logistic system that runs the Baofeng communication system. Introducing these radios has made it easier for the business to adapt to various needs and communicate more efficiently between employees. Therefore, the organization has managed to improve promptness, make resource allocation more efficient, and reduce delivery time. Besides, the organization's efficiency in daily operations increased, and customer satisfaction was also achieved.

6.3 COMMUNITY ENGAGEMENT

6.3.1 LEVERAGING ONLINE PLATFORMS

In the amateur radio community, online forums are dedicated to amateur radio, such as /r/amateurradio on Reddit and groups on Facebook, where professionals share their knowledge, collaborate to get things done, and engage the community. Members engage in discussions, share their experiences, and pass inventory to each other on the topic of the most advanced features of Baofeng Radio. For instance, members of this community build tutorials about programming Baofeng radios, solving the issues that may occur, and gathering information on antennas. These online tools enable amateur radio fans to grow their abilities, affiliate with others who share the same hobby, and add fresh ideas to the community's general knowledge.

6.3.2 EDUCATIONAL AND OUTREACH PROGRAMS

Not long after Baofeng radios debuted, they quickly became the most used tool in educational institutions across the country to introduce students to amateur radio communication and wireless networking. In a high school STEM project, Baofeng radios are integrated into the teaching; for example, to explain the propagation of radio waves and antenna theory to students, the most probable will be the communication protocols. Students take part in tasks that help them develop hands-on skills; for example, they design & program radios, carry out fieldwork, and participate in a contest amateur. This experience enables the students to develop profound perceptions of radio communication, generating their desire to learn about it practically.

Additionally, Baofeng radios have become key components of disaster management, the professional environment, and community outreach. For instance, communication is a priority in emergencies, e.g., natural disasters and public safety operations; thus, Baofeng radios are the optimal solution often when traditional communication networks collapse. In workplaces such as construction sites, factories, and warehouses, Vehenment radios provide simple intercommunication amongst the staff members, increasing efficiency and productivity. Or. The workers in places like factories, construction sites, and warehouses use Baofeng radios to communicate easily on the field, which helps save time and productivity. The organization of events and festivals and the cooperation of organizers and volunteers are facilitated within the community engagement activities through Baofeng radios.

The practical use of Baofeng radios and their cases reveal how differently the device can be handy across many sectors and various ventures. By offering stable communication channels, boosting collaboration, and endowing individuals, the Baofeng radios remain a crucial instrument for moving innovation, justness, and connection forward at a crossroads.

7. ADVANCED COMMUNICATION STRATEGIES

With the development of radio broadcasts, it becomes important to learn how to apply more advanced tactics to ensure effectiveness and also to cope with the intense competition in the radio communication field.

In this chapter, we will be going into detail about the best communication methods with the help of IT integration, counter-communications tactics, and long-range communication approaches. Taking advantage of these advanced approaches, you will get to the next level of communication competence, and you will be able to stay online and communicate in an instant in case of an emergency. Thus, whether you have been using radio for ages already or have just started with it, the following chapter has a lot of useful lessons and practical tips that will help you bring your skills to a new level and always stay in the loop on the latest developments in the field of radio.

7.1 DIGITAL AGE RADIO

7.1.1 AN INTEGRATION TO DIGITAL TECHNOLOGIES

The world of radio communications was amplified by the addition of digital modes, e.g., Digital Mobile Radio (DMR) and Digital Smart Technologies for Amateur Radio (D-STAR). Digital types have greatly surpassed the old analog system, providing listeners with remarkably better sound quality, outstanding data transmission capabilities, and more resistance to noise and interference.

Among the critical implications of digital types is the embeddedness of the Internet in the interaction of telecommunication and VoIP. For example, connecting radios to the Internet via Internet-based networks will lengthen the radio impact for interaction that crosses the traditional radio frequencies and enables global connectivity and real-time collaboration. Moreover, through internet-linked repeaters and reflectors, users will have the opportunity to set links between distant locations, thus fostering long-distance communication and interoperability between various radio systems.

Furthermore, digital technologies offer other sophisticated functions like text messaging, GPS tracking, and remote control options, some of which can increase amateur radio's effectiveness. These tools are helpful, especially in emergencies where the communication must be quick, and the information must be trustworthy to respond more efficiently and aid the distressed people.

Digital communication has entered the amateur radio area, dramatically and significantly impacting people's communication. Future breakthrough technologies like

SDR, AI, and decentralized networks are the building blocks for amateur radio tech development.

SDR technology provides excellent versatility and adaptability when designing radios, allowing for digital signal processing and modulation in a real-time environment. This new generation of technology brings features like adaptive modulation, spectrum sharing, or dynamic frequency allocation, enhancing the system's efficiency and flexibility.

AI-enhanced algorithms provide for radio performance tweaks and auto-adjusting parameters such as modulation, coding, and error correction to guarantee throughput and reliability. Moreover, AI-based predictive analytics can envision network jams and interference, consequently enabling proactive response mechanisms to be created to maintain the integrity of signals effectively.

Decentralized networks based on blockchain technology or mesh protocol eliminate single points of failure and enable resilience and redundancy by forming peer-to-peer communication without the requirements of a centralized infrastructure. These networks support people and communities in establishing more muscular communication systems in poorly served or remote places, efficiently promoting emergency communication, disaster response, and community building.

By adopting trends and innovations, amateur radio enthusiasts can maintain the leading edge of digital communications technology, locking many neunlocksns/new possibilities and opportunities for exploration and experimentation.

7.2 COVERT COMMUNICATIONS

7.2.1 TECHNIQUES FOR SECURE COMMUNICATIONS

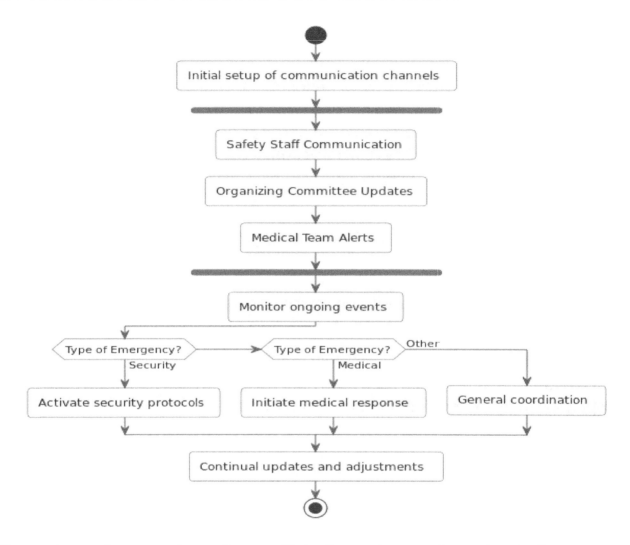

At times when privacy and security constitute the endgame, then clandestine communication approaches should be put to use. The secure communication methods comprise encryption, frequency hopping, spread spectrum modulation, and low-probability-of-intercept (LPI) techniques.

Encryption ensures zero exposure or interception of data being transmitted to unauthorized people. High-grade encryption methods, e.g., AEA (Advanced Encryption Standard) and RSA (Rivest–Shamir–Adleman), can effectively secure the client communication, thus protecting the data from unauthorized access as well as from any interference.

Frequency hopping involves a rapid change of transmission signal frequencies in a pseudorandom order, complicating the task of intercepting and jamming messages for adversaries. The spread spectrum modulation distributes the signal energy throughout a broader frequency band, decreases the odds of interference from a narrowband source, and boosts signal security in interception.

Such low-probability-of-intercept (LPI) methods decrease the probability of detecting transmissions by reducing signal power, using directional antennas, and using purposefully shaped waveforms. Covert communication methods, with their high interception rates, are critical to operational security and help to keep sensitive information hidden from the enemy.

Nevertheless, it is relevant to have legal and ethical implications for using secret communication techniques. Many jurisdictions regulate radio transmission encryption, and using covert communication methods without permission is considered illegal. Deviation from privacy laws or telecommunications regulations may bring a punishment. Users' tasks include observing laws and ethical norms, but there is a fine line between protecting information and observing the law or moral principles.

7.2.3. COVERT OPERATIONS CASE STUDIES

The numerous case studies present a deeper understanding of how covert communication operations play out in reality and how vital they are for executing a mission. During special forces operations, radios in various parts of the battlefield used encryption for commanders to steer the tactical movements and maintain security. Also, during police operations, the spies used covert communication to gather information by guessing, watching, ing and catching, catching, catching, catching, and revealing themselves.

7.3 LONG-RANGE COMMUNICATION TACTICS

7.3.1 MAXIMIZING RADIO REACH

Long-range communication involves antenna selection, positioning, and ambient noises, which must be considered. You must know which will give you the most significant potential for signal strength and clarity over the most extended distances. For instance, Yagi and log-periodic antennas direct waves, allowing for longer transmission ranges and better signal reception, making communication over large distances much more accessible.

Signal transmissions can be further increased by raising antennas at higher elevations, using rooftops or towers, as they usually have fewer obstructions and a lower degree of signal loss. To allow better orientation of transmitted and received antennas, clear line-of-sight paths are critical for maximizing signal propagation and minimizing the signal loss that buildings, trees, and terrain features could cause.

Along with this, environmental features like atmosphere conditions, landscape topology, and electromagnetic interference can also be responsible for the performance of long-range communication. After defining these factors and taking measures such as adjusting transmit power, modulation scheme, and antenna polarization, one can enhance communication reliability and effectiveness.

7.3.2. EXPERT TIPS AND TECHNIQUES

Carrying out aerial surveys, mounting antennas in the best sites, and evaluating transmission paths.

Applying propagation prediction tools and repeaters to recreate signal propagation and identify coverage areas.

Attempting various solutions, such as stacked and phased arrays, is the best solution for signal strength and directionality improvement.

Using signal amplifiers, preamplifiers, and signal boosters when lobbing and extending the transmission range inhibits the corruption of the signal.

The operator must cooperate with other operators and participate in DX (long-distance) contests and events to get the best result in long-range communications. Incorporating these expert tips and tricks will enable amateur radio fans to communicate more in the range beyond the radio frequency limit.

7.4 CASE STUDIES

COMMUNICATION STANDARDS IN RESPONSES TO EMERGENCIES.

Generally, communication systems (i.e., television and radio networks) are not spared in a natural disaster (think hurricanes and earthquakes), especially during the initial stage, which may hinder communications systems. Situations like these are often prone to disconnection as the digital devices' communication protocols are invaluable for maintaining connectivity and coordinating the rescue process.

One of the explicit examples is the post-Hurricane Katrina experience; when the committed operators of the amateur radio started PSK31 and Winlink digital modes, their successful communication network distress availed where other means were failing. By implementing these digital protocols, operators could send messages, provide the required information, and organize rescue operations, though there was widespread infrastructural damage.

Employing digital communication procedures that bypass the communication hindrances aided responders in furnishing emergency assistance to communities struck by the disaster. This study, in particular, shows the persistence and dexterity of information technologies in urgent situations, thus advising their usage in the emergency response framework.

COVERT COMMUNICATION IN LAW ENFORCEMENT ACTIVITY OPERATIONS

In law enforcement, the need for covert communication lines is vital because one of the key things is to achieve operational security and the prevention of the disclosure of sensitive things. Whether they use encrypted devices or elaborate means of cover, covert communication skills ensure that agents can find a way to communicate safely, even under the prying eyes of enemies and criminals.

As an illustration, in a joint investigation concerning organized criminal networks, law officials resort to high-tech gadgets, including widely encrypted and frequency hopping spread spectrum (FHSS) digital radios. Thus, by providing an FHSS, radio transmitters using high and fast switching rates made it impossible for any enemy to scan and decode them in the future.

This kind of secure, cryptic, two-way communication allowed the agents to coordinate with surveillance activities minus the possible alerting of the target. With this in mind, police departments came across valuable information, and their efforts helped arrest high-profile and organized criminals, which were instrumental in the undermining of criminal syndicates.

7.5 HOW TO CREATE A SECURE COMMUNICATION NETWORK

Here's a step-by-step guide on how to create a secure communication network:

CHOOSE SECURE COMMUNICATION PROTOCOLS

- Administer a range of encryption IT protocols, including AES and DESE, to fortify your communication channels.
- Run your radios so that they can utilize secure modes like DMR (Digital Mobile Radio) or Tetra, which entails encryption capabilities for safe transmission.

FREQUENCY HOPPING SPREAD SPECTRUM (FHSS) SETUP

- Introduce the FHSS feature on your radio to improve your communication security, as it is fast and helps to switch frequencies.
- Tuning the rate of jumping and the range of the frequency allows for avoiding detection and jamming.

IMPLEMENT AUTHENTICATION MECHANISMS

- Roughly configure SACS (Secure Access Control System) or RAP (Radio Authentication Protocol) to exchange your traffic pods with each other so that communication endpoints can be identified.
- Implement a two-factor authentication with unique keys or passwords to any outside connections for your network to minimize the risk of unauthorized access.

CREATE SECURE COMMUNICATION CHANNELS

- Establish secure communication channels, separate from public yet non-secure channels, to avoid leakage of any sensitive information.
- Allot each channel a unique channel ID and encryption key to keep hackers at bay.

TEST AND VALIDATE NETWORK SECURITY

- Carry out comprehensive testing of your secure network architectural setup so that the mechanisms of encryption and authentication can be seen and if they are functioning correctly.
- Put encryption testing tools and spectrum analyzers in place to confirm that your chosen techniques work as they should.

MONITOR AND MANAGE NETWORK TRAFFIC

- Installing network monitoring solutions to keep trails of communication flow in an actual-time manner.
- Oversee unauthorized access attempts and monitor for suspicious activity by promptly responding and taking appropriate measures to ensure security.

7.6 TROUBLESHOOTING TIPS FOR CREATING AND MAINTAINING A SECURE CONNECTION NETWORK

SIGNAL INTERFERENCE

Issue: Signal interference can jam radio communications and reverse the intended-to-be-secret messages.

Troubleshooting Tip: Identify the causes of the disturbances, such as the nearby electronic devices, or environmental variables, such as the slopes or weather conditions. Changing antenna position or tinkering with frequency settings to mitigate interference can ensure signal quality and integrity.

AUTHENTICATION FAILURES

Issue: Authenticating failures may indicate an access attempt by an unauthorized person or may be due to misconfigured security items.

Troubleshooting Tip: Check and confirm credentials and ensure that they are also available on both communicating points. Examine authentication processes and the used keys to identify if any mistakes were made or keys were obtained without authorizations.

NETWORK CONGESTION

Issue: Network congestion causes delay or packet loss, which causes communication failure, as in the case of delays or lost packets.

Troubleshooting Tip: Watch for the network traffic and see if there are crowded elements present or if there are choke points (for communication paths). For the purpose of priority communications, the QoS (Quality of Service) mechanism can be used to address congestion and make flow fluent.

ENCRYPTION KEY MANAGEMENT

Issue: Tightening key management practices in poor encryption can expose system security loopholes or compromise communication.

Troubleshooting Tip: Updated encryption keys and rotation should be done in a timely manner to prevent other parties from breaking the code. Implement safe key distribution schemes and remain compliant with encryption best practices above others in order to ensure confidentiality.

HARDWARE MALFUNCTIONS

Issue: Sometimes hardware spoilages or fallout of equipment can halt communication and hazard network security

Troubleshooting Tip: Conducting periodic installation tests on hardware components such as radios, antennas, and network infrastructure will be part of your routine duties. Rushing to get faulty or old equipment replaced contributes not just to improving the stability and security of the network but also to the overall business bottom line.

SECURITY BREACHES

Issue: Data leakage, system intrusion and interruption, and other unauthorized activities undermine the security and confidentiality of the network.

Troubleshooting Tip: Keeping an eye on network activity for signs of fraudulent activity or scan attempts for unauthorized access. Deploy ID systems and data security approaches in order to identify and deal with possible intrusions quickly.

ENVIRONMENTAL FACTORS

Issue: Environmental factors, including electromagnetic interference or adverse weather phenomena, should be considered while planning for communication capabilities.

Troubleshooting Tip: gauge the environmental hazards and how they influence communication performance. Adopt security measures like protecting equipment from electromagnetic interference or putting a housing shelter on the front line space to eliminate the risks caused by environmental conditions.

SPECIAL CONTENT 1: EMERGENCY COMMUNICATION CHECKLIST

👇 **SCAN HERE TO DOWNLOAD FOR FREE** 👇

SPECIAL CONTENT 2: THE ULTIMATE HAM RADIO GUIDE

👇 SCAN HERE TO DOWNLOAD FOR FREE 👇

8. THE TECHNICAL DEEP DIVE

Whether you're a newbie or a pro, this chapter will ensue you'll get everything out of your radio receiver. We'll begin with a detailed analysis of the rules of radio waves, including different types of waves, frequencies, wavelengths, and propagation. We'll cover the technical details of Baofeng radios next; their power output, modulation, bandwidth, and sensitivity will all be addressed. Further, we will look at different types of antennas and their working and how to choose the one suitable for you. You'll also understand the usual problems associated with Baofeng radios, how to detect and rectify them, and troubleshoot routine maintenance to help the device last longer. By the end of this chapter, you'll be able to grasp the technical aspects of Baofeng radios and, thus, will be able to use all the basics it offers to your full access.

8.1 SIGNAL THEORY AND COMMUNICATION

8.1.1 BASICS OF RADIO WAVES

Before we start to show the technical aspects of radio communication, it's crucial to know the essential characteristics of radio waves. Radio waves are electromagnetic waves that can be described in terms of the frequency, wavelength, and amplitude of their waves. The wavelength is the distance between successive crests or a trough, and the frequency is the number of oscillations occurring in a second. However, intensity is the degree to which the wave is strong or not.

On this account, let's note that radio waves are also used for applications like broadcasting, satellite communication, and wifi. The spread of radio waves is an essential tool in designing and implementing communications and networks. This procedure also includes recognizing influences that might affect signal propagation, like the terrain, atmospheric conditions, and frequency bands. Through investigating radio wave propagation, we guarantee a satisfactory communication system that guarantees reliable transmission under various situations.

Key Concepts:

Wavelength and Frequency: Wavelength and frequency characterize radio waves, which have fundamental properties. A case of anti-totality is consequently noted, i.e., the higher the frequency gets, the shorter the associated wave gets, and the vice versa situation goes. The higher frequencies equate to the lower wavelengths by contrast to the lower frequencies, which stand for the longer wavelengths. Knowing this connection is critical to grasping how radio waves travel through different media in different modes and determining radio waves' reflection on other objects.

Propagation Modes: Through various propagation modes, patterns stand for specific mechanisms and conditions. Three primary propagation modes include: Three primary propagation modes include:

Ground Wave: Ground wave propagation is an occurrence when waves fulfill the retarding force of the Earth by going along its curvature. This mode is mainly used for short-range communications where even the AM broadcast band is included, but the utilization is more for short-range frequencies.

Sky Wave: Propagation of the sky wave is the phenomenon in which the ionosphere refracts the radio waves and bounces them back to Earth, facilitating line-of-sight transmission of radio signals to a longer distance. Such a method is essential, as it works with HF communication mode, allowing a global reach under particular atmospheric expositions.

Line-of-Sight Propagation: Unlike other radio waves that travel in straight lines from a transmitter to the receiver through a line-of-sight path, line-of-sight propagation occurs where there are no significant obstacles or reflections. This mode is critical for places not far away, provided that the signal line from the transmitter can be followed to the receiver.

Every transmission type requires a different set of characteristics, and limitations have to be understood to improve the overall performance and use the optimal frequency range and the transmission methods.

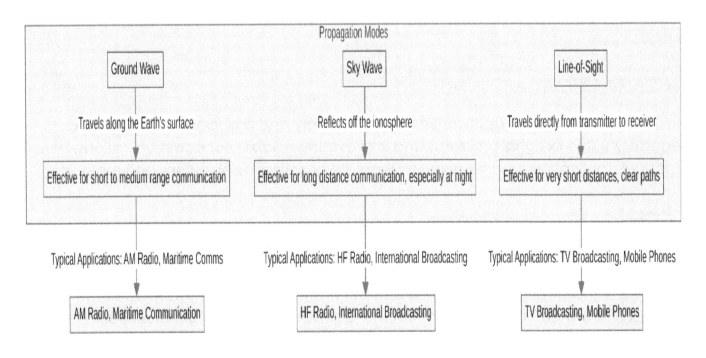

FACTORS AFFECTING PROPAGATION

A couple of components determining the desired signal transmission can cause inaccuracies and loss of communication in faraway places. Some of the key factors include: Some of the key factors include:

Atmospheric Conditions: The characteristics of the atmosphere, e.g., temperature, humidity, and pressure, play a part in the restrictive factors in the movement of radio waves, which are the density and composition of the atmosphere. An instance illustrates the heating temperature, highlighting the atmospheric layers refracting or absorbing the radio waves differently.

Terrain Topology: Due to hills and valleys around, as well as complicated built-up areas, the radio waves can be reflected, diffracted, and even shadowed by the terrain. We must be able to foresee the terrain nodal of the area to get the signal coverage prediction and antenna equipment placement.

Electromagnetic Interference (EMI): EMI from external sources, like power lines, electronic gadgets, and atmospheric noise, can pollute radio transmissions by creating undesirable signals and errors. This leads to issues of poor quality and reliability with communication. Phase inequality is one of the critical EMI factors in signal integrity. Reducing the EMI through shielding, filtering, and frequency management is necessary for high-quality and stable signal transmission.

Whether it is about approaches adapted to the environment for their optimization, principles of communication system working, or methods of radio system management, all these will be clear and comprehensive due to extensive knowledge gained on that level.

8.1.2 PRACTICAL APPLICATIONS

Mastering the real-world applications of signal theory and propagation becomes a necessary skill to fix signal problems and improve the reliability of communications. Here're practical applications and techniques: Here are practical applications and methods:

SIGNAL STRENGTH MEASUREMENT

When reading the data, the Received Signal Strength Indicator (RSSI) is used to tell how strong the signals received are, so understanding this is essential. RSSI, the acronym for Received Signal Strength Indication, is a meter that almost every modern radio device has and simply measures the strength of the received signal. As a general rule, the RSSI sign is typically shown in dBm (decibels relative to 1 milliwatt), and a higher RSSI number means a stronger signal, while a lower RSSI number means a weaker signal. RSSI levels should be monitored to determine signal quality while corresponding to the appropriate decisions regarding communication.

ANTENNA SELECTION AND POSITIONING

Determining the necessary type of antenna featuring an adequate placement issue is essential for controlling signal strength and interference. Different antennas have unique functions and are usually tailored for particular situations. For instance, a directional antenna is more of a rotary widget that transmits signals more accurately in one specific direction and, hence, may be used to communicate over long distances from point to point. On the one hand, because the omnidirectional antennas broadcast the signal uniformly in all directions, they are helpful for general coverage applications.

Height, orientation, and surroundings are to be specified for effective positioning. Antennas should be positioned at a high place to avoid obstructions and allow direct connection. Furthermore, antennas should be installed to separate them from interference sources, including power lines and electronic gadgets, to eliminate the effect of signal attenuation.

INTERFERENCE MITIGATION

The signal can be hindered by external sources, which can degrade its quality and disrupt communication. Methods for interference minimization are instrumental factors in preserving the integrity of the transmission process. Most of the time, interference is probably from the proximity of electronic devices, atmospheric sound, and co-signal.

To mitigate interference, operators can take several measures:

Frequency Management: Joint stations may select the frequency with minimal interference; this, in conjunction with frequency coordination techniques, is a surety in mitigating co-channel interference.

Shielding: Shielding fragile apparatus and leading against magnetized radiations into the environment serves to lessen the level of transmutation of signals.

Noise Filtering: Putting in place quietening gadgets and suppressors will decrease the transmission of undesired messages as well as the background noise, thus leading to signal improvement.

Distance: Increasing the meaning of the radio with those potentially from the devices of effects by interferences on signal quality is a possible way to handle this.

Antenna Placement: The antennas should be positioned in an area away from the sources of interference, and the antenna orientation should be opted to minimize the sign interference due to electromagnetic interference.

Network administrators can use these interference mitigation methods to get better signal quality, saturate communication reliability, and promote the strong performance of radio systems in all kinds of environments.

8.2 ANTENNA THEORY

To fully grasp advanced antenna design concepts, it's essential to understand the fundamental principles of antenna theory. We will cover topics such as radiation patterns, impedance matching, and polarization, providing readers with a solid foundation for exploring more complex antenna designs.

Radiation Patterns: Antennas emit electromagnetic waves in specific patterns determined by their geometry and configuration. We will discuss different types of radiation patterns, including omnidirectional, directional, and sectoral patterns, and their implications for signal coverage and propagation.

Impedance Matching: Achieving impedance matching between the antenna and the transmission line is crucial for maximizing power transfer and minimizing signal reflection. We will explain the concept of impedance matching and techniques for tuning antenna impedance to match that of the transmission line.

Polarization: Antenna polarization refers to the orientation of the electromagnetic wave's electric field with respect to the Earth's surface. We will explore the significance of polarization in radio communication and its impact on signal propagation, highlighting techniques for optimizing antenna polarization for specific applications.

8.2.1 CUSTOMIZING THE ANTENNA FOR SPECIFIC PURPOSES

All communication scenarios require different configurations and, therefore, tailor-made designs of specialized antennas according to the required application. Here are the different types:

DIRECTIONAL ANTENNAS

Directional antennas, e.g., the Yagi-Uda antenna and parabolic antenna, are essential to radio communication. Their unique ability to concentrate radio waves in a particular direction has made them an invaluable part of various communication systems. Rather than spreading the signal energy over a broad area, this narrow radiation beam provides far more signal strength and gain, and thus, the signal can be achieved over distances. The second part will deal with the design principles as well as construction techniques of a directional antenna and illustrate these with some practical examples, such as how it was used in the point-to-point link and satellite communication services.

Yagi-Uda Antennas: Yagi-Uda antennas, also called Yagi antennae, have various parallel elements stretched along a boom and a driven element that is excited with the transmission or the reception. Various features like reflectors and directors help shape the pattern to concentrate energy on a specific area. By using the selectable length and spacing of these components, engineers can figure out the most efficient design for the communication distance and target frequency.

Constructing a successful Yagi-Uda antenna requires obeying all specified parameters strictly, as the dimensions and spacing of the elements are decisive for the antenna's operating characteristics. Engineers rely on mathematical models and computer simulations, instruments they use to create the most appropriate antenna designs. The most common construction materials include aluminum tubes or wire for the actual UAS and a non-conductive boom to bear them. The construction of the antenna takes place next, which then proceeds to the tuning and testing to improve the performance level.

Parabolic Antennas: Parabolic antennas, also known as dish antennas, make use of parabolic reflectors, such as a satellite dish, to concentrate incoming radio waves on a detection device, for example, a dipole or horn antenna. The factor that makes the reflector curved is that it ensures that all the incoming waves bounce toward the focus point where the horn takes the signal. This antenna model provides you with wide bandwidth and beam shaping, which makes telescopes ideal for long-range communication and satellite tracking.

Directional antennas get broad applications in different means of communication. In point-to-point links, antennas of the Yagi-Uda type are preferably used to reach long distances in areas where broadband networks are not available. Such a distance could be relevant in rural broadband networks and amateur radio communication. Parabolic antennas have become irreplaceable tools for satellite communication, and they function as receivers for satellite signals or transmit data to spacecraft in an accurate and efficient manner.

OMNIDIRECTIONAL ANTENNAS

When using Omnidirectional antennas, radio waves are emitted uniformly on all sides, thus providing the surrounding area coverage needed for 360° communications. In this section, we will discuss the design specifications, optimization options, and use of fron- back antennas, such as vertical monopoles and terrestrial antennas. Designing Omnidirectional antennas includes diligent design and assembly to get the best performance and coverage. Here are the basic steps involved in constructing omnidirectional antennas: Here are the basic steps involved in constructing omnidirectional antennas:

DESIGN PHASE

Determine the frequency range: Identify the working frequency band in which the antenna is going to operate; the functional parameter will be dependent upon that.

Choose the antenna type: Choose the ratio of wide-angle antenna, which will be suitable depending on the application and needed coverage area. The two general types you find are vertical monopoles and ground-plane antennas.

Calculate dimensions: Utilize antenna design software or mathematical formulas to arrive at the size of the antenna elements, which include the length of the radiating element and ground plane elements.

Determine materials: Match the selected materials for constructing the monopole antenna elements and the support structures. Aluminum or copper materials are the typical materials used for antennas' elements, and non-conductive materials for support structures, such as PVC or fiberglass, are also possible.

CONSTRUCTION PHASE

Prepare materials: Mark the antenna elements and supporting structure using a recognized marking method such as Saferank. Next, cut the parts accordingly using suitable tools like a hacksaw or pipe cutter.

Assemble the elements: Adhering the radiating element (vertical element) to the support structure, where it will be positioned perpendicularly. For planar grounding, the ground plane elements shall be almost parallel to the horizontal and not in the direction of the radiating element.

Mount connectors: Put connectors of the types SMA or BNC to the ends of the antenna parts in order to connect to the coaxial cable.

Test fit: Collect the antenna components and do a shake-down test to ascertain from device dimensions and item spacing that everything is aligned.

TUNING AND TESTING PHASE:

Adjust element lengths: Tune the sizes of the antenna elements remarkably to match the operating frequency in question. These could be prototypes for which some parts, like the ones for attaching the antenna, are trimmed or softened in length, and then they are tested, which is the performance of the antenna.

Measure SWR: Analyze the ratio of forward and reflected waves between the antenna and radio using a standing wave ratio (SWR) meter. Small element lengths or positions alter to eliminate SWR and enhance the antenna efficiency.

Conduct field tests: Set up the antenna in its given position and do field trials to test performance standards in real-world environments. Identify the signal strength and the coverage area for a network to make sure it works according to the needed requirements.

OPTIMIZATION PHASE:

Optimize mounting location: Find out the optimum post mount height and location through trial and error to deliver the best signal quality and reduce interference.

Fine-tune element positions: Wringing the last drop out of performance can be achieved by making tiny adjustments to the angles of the antenna elements if needed.

Consider environmental factors: Consider the environmental factors, such as natural obstacles and RF interference, when formulating an antenna system for maximized performance.

TROUBLESHOOTING COMMON SIGNAL ISSUES

Static and Noise: It is important to note that static and noise are dependable problems affecting communication quality during radio transmission. The reduction of static and noise is only possible through eliminating the sources of origin to apply successful mitigation options. Electromagnetic interference, atmospheric phenomena, and signal reflections commonly lead to the so-called static or noise genes.

To minimize the impact of static and noise, consider the following strategies:

Antenna Placement: Ensure that the antenna is placed at places that have minimum interactions with electrical interference signals, including power lines, electronic gadgets, and fluorescent lights. The issue of optimal antenna placement will certainly minimize the possibility of the radio being strained by external noise sources.

Shielding: Apply walls with shielding materials to safeguard private components and cables from getting affected by electromagnetic interference. The panel can also reduce the amount of noise shouting to be detected and thus enhance the quality of the communication.

Noise Filtering: The in-flight state of affairs shall be controlled by noise filters and suppressors, which will reduce unwanted signals. It will provide the required rank and calmness, which will enable flying. This will imply a reduction of atmospheric noise.

Advanced noise filters that can boost signal clarity or reduce non-interference static are now available in the market.

Frequency Selection: Select frequencies that cause minimum interference and noise to get the highest possible information quality. By carrying out a range scan to reveal unoccupied channels, one can avoid the complications of co-channel interference and congestion of signals.

Grounding: Proper grounding of the radio equipment is a priority. Hence, it will be minimized, and electrical interference shall be avoided through ground loops. The grounding products correlate with the control of undesired electrical currents, allowing them to disperse and prevent the interference caused by static and noise.

SIGNAL FADING:

The second issue of poor signal quality is related to the signal fading effects. Signal strength fluctuations characterize the signal fading as a function of time and distance. These variations are usually caused by changing weather conditions, significant obstacles like hills and buildings, and multipath propagation. The fact that signal fading can be enhanced by compensating techniques is equally essential for maintaining the reliability of communication channels exposed to dynamic environmental conditions.

To address signal fading, consider the following techniques:

Diversity Reception: Many antennae or antenna elements have been used in the process, and these elements can be deployed around different angles and polarization orientations to capture signals coming from various directions. Diversity reception is for a more robust signal by adding the received signals through propagation paths, and this, in turn, reduces the problems of signal fading.

Automatic Gain Control (AGC): Employ AGC circuits to tune the gain dynamically in the receiver such that the gain gets adjusted in response to the power level of the signal. AGC plays an essential role in keeping the sound space relatively steady, thus minimizing the likelihood that the listeners will notice any variations in the levels of the susceptible signal, such as fading.

Forward Error Correction (FEC): Design FEC algorithms to correct errors that result from poor signals and additive white noise. The FEC tactics ensure the robot convoy operations, enabling the recovery of lost or corrupted information, hence improving communication reliability.

Adaptive Modulation: Convey through adaptive modulation with management techniques that alter the modulation scheme and transmission parameters toward the channel conditions. Adaptive modulation can optimize bandwidth utilization and improve the performance of communication systems in frequency-selective environments through the dynamic sending of signals by varying conditions.

By using the given pragmatic methods, you'll be capable of overcoming the most widespread signal malfunctions, interference from static and noise, and signal weakness, which, on the whole, will improve Baofeng's radio effectiveness in the most demanding locations.

8.3 TECHNICAL TROUBLESHOOTING GUIDE

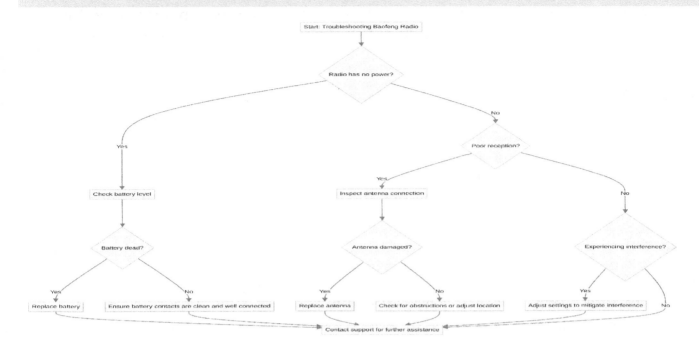

When your Baofeng radio develops defects, diving into the problem systematically is the best way you will be more effective in diagnosing and fixing them. Let's delve into diagnostic tools and techniques for troubleshooting Baofeng radios: Let's delve into diagnostic tools and techniques for troubleshooting Baofeng radios:

COMMON ISSUES:

No Power: Exhausted by many radio users, power is one of the most burning problems of radio users. To address this, it's essential to identify potential causes such as: To address this, it's necessary to identify potential causes such as:

Battery Issues: Identify possible wear and tear, discharging, and charging issues by thoroughly examining the corresponding parts of the battery. Use a discharge end to change the battery or use a recharge function if possible.

Faulty Connections: Carry out all the connections like battery terminals, power cables, and connectors for tightness, leakiness, corrosion, or damage. Make sure that the power of the reconnection is restored.

Internal Component Failure: If everything from the external to the battery is in good condition, the potential bad guys are likely caused by internal component failure. Have a tech guru interested in the problem run a diagnostic test and see the latest upgrade.

POOR RECEPTION:

The main problem that needs to be resolved is poor reception, which has a negative impact on communication quality. To troubleshoot poor reception, consider potential causes such as:

Antenna Problems: Inspect the antenna for any visible defects, loose connection, or poor alignment. If the audio sounds weak, try replacing or adjusting the antenna for better quality.

Signal Interference: Find out the sources of interference, for instance, satellites and other astronauts, nearby electronic devices, and the nature of co-channel interference. The system's efficiency is enhanced by reducing or terminating the inputs that impede reception.

Receiver Sensitivity Issues: Sounds of received signals may be falsified or turned garbled due to the radio's receiver's sensitivity in problematic cases. A technician should be consulted for diagnostic issues; some receivers can be fixed with good feedback.

DIAGNOSTIC TOOLS

Multimeter: The multimeter is a remarkable utility tool employed to detect diverse electrical parameters, including voltage, continuity, and resistance. To diagnose power-related issues:

Measure Battery Voltage: Measure the voltage using a multimeter to measure the battery's output. The results from the electrical meter can be too low, indicating a dying or faulty battery.

Check Continuity: The interruption of the flow of electricity can be anticipated by switching on the continuity across the power cables and the connectors. The drop of the latch might suggest a fact of a disconnected wire.

SWR METER:

An SWR meter is one of the most essential devices for evaluating antenna performance and discovering impedance problems. It includes the exact length of the antenna compared to the size of the radial and coax cable impedance. To diagnose antenna-related issues:

Measure SWR: Link an SWR meter between the radio and its antenna port to measure signal amplitude. Analyze the impedances via SWR measurement and thwart possible impedance mismatches.

Adjust Antenna: Through the reading of SW R, adjust the length of the antenna or change the positioning to ensure that the SWR values have been maximized, thus improving the reception.

SPECTRUM ANALYZER:

Spectral analyzers are highly complex and intelligent diagnostics that measure signal spectra and detect noise that causes interference or distortion. To diagnose signal-related issues:

Analyze Signal Spectrum: Attach the benchtop spectrum analyzer to the transceiver's output port and observe the signal spectrum displays. Spot any peaks, spikes, or abnormalities that can be attributed to media noise and interference.

Locate Interference Sources: Use spectrum analyzer output to isolate the interferences in the vicinity, particularly electromagnetic waves from another transmitting station, nearby sources of electronic noise, or atmospheric radio interference. Take precautions to minimize clashing and enhance signaling reception.

These diagnostic methods and tools can quickly pinpoint and tackle your BaoFeng radio problems, thereby providing optimum performance and extreme reliability in all conditions.

8.3.2 MAINTENANCE AND REPAIR

As a Baofeng radio user, remember that routine maintenance and occasional repair are vital in maintaining a long-standing unit with superior performance. Let's explore some essential maintenance tasks and repair techniques that you can perform yourself:

ROUTINE MAINTENANCE

Cleaning: Often, dust, dirt, and debris trapped in the external and internal mechanism parts are removed to ensure unhindered communication and maintain its maximum performance. Use a mildly soft and dry cloth to wash off the outside surfaces and swab them to remove dirt particles or dust from the controls, buttons, and ports. For dirtier spots or buildup, you can lightly dip the cloth into water instead of wetting it. Carefully avoid letting the moisture get to the radio's interior.

Battery Care: Battery management is vital to have the most extended battery life and an uninterrupted power supply for this radio. Follow these battery care practices:

Avoid Overcharging: Don't connect the charger to the battery after charging for too long. The battery lifespan will drastically reduce if the charger is overcharged.

Proper Storage: Please avoid letting the battery stay uncharged in too-hot or cold environments or areas with direct sunlight, or it may become degraded.

Regular Use: Make sure the battery is periodically discharged and recharged to ensure that it is working at an optimum level and also to avoid unexpected capacity loss due to the prolonged inactivity of the battery.

FIRMWARE UPDATES

It is essential to keep this device updated with the new firmware versions to access new features, improvements in performance, and bugs that are fixed for your Baofeng handheld radio device. To perform firmware updates:

Check for Updates: Visit baofeng.com or any authorized reseller to get the latest firmware versions for your radio model.

Follow Instructions: Please follow the manufacturer's requirements for not disrupting the radio while changing the firmware and ensuring the radio is constantly connected to the electricity source.

SIMPLE REPAIRS

Antenna Replacement: Replacing an antenna becomes vital to meet regular communication by knowing how to use a compatible replacement for the old-fashioned one. To replace the antenna:

Remove the old antenna by unplugging it from the radio antenna connector.

Screw the new antenna onto the antenna connector until it's fixed. You can even test it to see if it is in place correctly.

SPEAKER REPLACEMENT

Bearing in mind the situation when the speaker in your Pozyx radio gives any trouble, replacing it can easily contribute to the quality and clarity of the audio. To replace the speaker:

Open the Radio case to get a hold of the internal parts.

Disconnect the wires connected to the old speaker and place it from the mounting position.

Mount the speaker you have brought by twisting in the wires yourself and using screws or adhesive to secure it.

BUTTON OR KNOB REPAIR

If the vase on your radio becomes cracked or unresponsive, replacement leads to restoring functionality and usability. To repair buttons or knobs:

With all due care, unscrew the buttons or knobs that are broken out of the radio housing.

Take a look at the button or knob contacts and at the contacts, where they connect to the circuit board, clean them of any dirt or debris.

If feasible, replace the broken button/knob with a fresh one and put the casing back on the radio.

You can perform the daily maintenance responsibilities like an easy remedy to prolong the radio's lifetime and always have your radio reliable.

8.4 SOFTWARE AND FIRMWARE UPDATES

8.4.1 KEEPING YOUR RADIO UPDATED

Software and firmware updates are vital as they keep the Baofeng radio up-to-date with the latest features, improvements, and security patches. undefined

IMPORTANCE OF UPDATES

Recognize software and firmware update functions to fix software bugs, vulnerabilities, and compatibility issues; these updates can add new features and functionality.

PERFORMING UPDATES

Official Sources

Take updates from authentic Baofeng sources, e.g., the Baofeng official website or certified reseller, to signify trustworthiness and safety.

UPDATE PROCEDURE

Follow the manufacturer's recommendations to update the software and firmware using the proper installation and configuration process.

8.3.2 CUSTOMIZING SOFTWARE FEATURES

Users can also customize the firmware to provide advanced functionalities and more usability. undefined

THIRD-PARTY SOFTWARE

CHIRP: Employ CHIRP third-party software to program your Baofeng radio, customize channel settings, and manage command data more effectively.

Firmware Mods: Studies the firmware code modifications done in the amateur radio community and adds new features and functionalities to your radio.

USER SETTINGS

Channel Organization: Customize your radio station's names, frequencies, and channel settings to arrange them in a way that is easy to navigate and access.

DISPLAY SETTINGS:

Adjust the display settings like backlight brightness, contrast, and font size to make text visible and accessible to read under varied lighting conditions.

By changing the software of your Baofeng radio, you can personalize it to meet your unique requirements and your liking, which, in turn, boosts usability and performance.

To sum up, learning the technical details of Baofeng radios requires you to be a combination of a theory learner, a skillful practitioner, and an experienced tester. Knowledge of signals theory and propagation, diagnostic tools and techniques, routine maintenance and repairs, and software and firmware updates ensure you have the skill set necessary to optimize the usability and reliability of your Baofeng radio for an extended period.

8.5 EXPERT INSIGHTS

The following are advanced tips from radio and communication experts on improving radio communication:

- Utilize advanced radio propagation models and measurements for emerging wireless personal communication systems, taking into account both indoor and microcell propagation environments.
- Implement cell splitting in cellular radio systems to increase capacity and accommodate more users by dividing cells into smaller cells with lower transmitter power and antennas with down-tilted radiation patterns.
- Take advantage of the upcoming FCC auction of 160 MHz of spectrum for PCS in the 1.8 to 2.2 GHz band, as well as the availability of ISM bands for a wide range of voice and data communication services.
- Design small wireless devices that are inexpensive and easy to use to meet the demand for ubiquitous access to wireless voice and data communication.
- Address new propagation issues in the unlicensed ISM band to design and install systems and evaluate their performance in a noisy environment.
- Consider the use of PCS, which differs from cellular systems by providing a range of telecommunication services using low power base transmitters and smaller cell sizes in the 1.8-2.2 GHz band, including two-way calling, high bandwidth data, voice, and video transmission services.
- Explore the potential of W-LAN for networking fixed and portable computers via a wireless data link in various locations, providing flexible and reconfigurable connectivity.

9. INTEGRATING BAOFENG WITH MODERN TECHNOLOGY

Baofeng radios offer a broad set of features usually needed in modern devices. They handle communications between different platforms, thereby providing efficient and smooth integration. Let's explore how Baofeng radios can be connected with smartphones, computers, and IoT devices: Let's explore how Baofeng radios can be connected with smartphones, computers, and IoT devices:

9.1 BRIDGING TECHNOLOGIES

Linking Baofeng radios with cell phones, computers, and the Internet of Things(IoT) creates better communication and collaboration opportunities. Here's a detailed guide on how to bridge Baofeng radios with various technologies:

Smartphone Integration: Different ways exist to bridge the gap between Baofeng radio systems and mobile phones. Please use Bluetooth adapters, RCA and FM receiver cables, and dedicated applications to allow users to remotely operate their radios, send text messages, and access GPS data for location quests.

Computer Connectivity: Baofeng handhelds can be hooked to computers via USB programming cables. CHIRP, or manufacturer-offered programmable software, works well with these models. This feature simplifies radio programming, software updates, and data exchange between the radio and computer, which have become more accessible, and access to these functions has become direct.

IoT Integration: Make BAO FENG radios smarter by embedding them with IoT devices to improve automation capabilities and remote monitoring functions. With Ti and IoT Platform integration, users can control devices remotely, receive alerts, and take sensor readings in different environments.

COMPARATIVE ANALYSIS OF INTEGRATION TECHNOLOGIES

We will focus on the comparison of different integrating techniques, including Bluetooth, Wi-Fi, and cellular, in the intelligibility of the radio technology, Baofeng. The technology avails different capabilities and quarrels, each of which has its own strengths and weaknesses.

BLUETOOTH INTEGRATION:

With Bluetooth technology, short-range wireless communication between all devices can be effectuated, which makes it perfect for applications that need the closeness of devices. Feng radios can be applied by smart devices such as smartphones or tablets via Bluetooth connection that allows for remote control, data transfer, and hands-free operation. Integrating Bluetooth brings low power consumption and ease of configuration, which are characteristics that make it widely used in cell phones and other portable devices. Conversely, Bluetooth might be ineffective for long-distance communication due to its small radio range.

WI-FI INTEGRATION:

Compared to Bluetooth technology, Wi-Fi offers the advantage of providing high-speed wireless connectivity over greater distances, which makes it ideal for those applications that require a range of connections. By manufacturing Baofeng radios with Wi-Fi network capabilities, there will be additional functions such as remote surfing, firmware updating, and transmitting data. Wi-Fi integration can provide a faster data transfer rate and a longer range than Bluetooth. Thus, it can be used for real-time streaming, telemetry, and remote wireless control. Nevertheless, this might involve a more sophisticated initial setup and configuration on the side of Wi-Fi, which could be more complicated compared to Bluetooth.

CELLULAR CONNECTIVITY:

The cellular link enables Baofeng radios to interface with telecommunication channels and consequently gain online communication services and internet functionalities. This is where the power of baofeng radios increases; you can make calls, send text messages, and share data from anywhere with cellular coverage. The option for cellular technology in integration is beneficial, especially in operations that demand long-distance communication or remote control, thanks to its capability of providing wide signal coverage and reliability. On the other hand, the cellular integration might include additional expenses like subscription payments and data charges, which could be required by the cellular service provider.

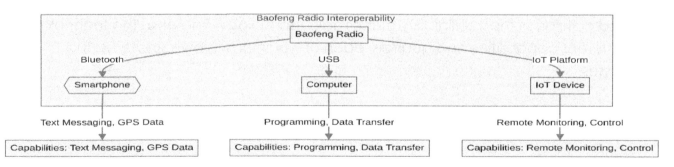

PRACTICAL USE CASES

The prime advantage of this integration of Baofeng radios with current devices is that they have multiple uses in all sectors of society and industry. Here are some practical use cases demonstrating the advantages of interoperability: Here are some practical use cases demonstrating the advantages of interoperability:

Emergency Response: The merging of smartphone apps and Baofeng two-way radios with smartphone apps offers smooth communication for emergency personnel during emergencies. GPS functionality through built-in location tracking and text messaging, especially among team members synchronization and information distribution, have a big contribution.

Remote Monitoring: Utilization of a connection between Baofeng radios and IoT sensors allows the surveillance of environmental problems, equipment conditions, and security systems at a distance. Real-time data transmission enables effective decision-making to solve an urgent problem, even in difficult circumstances, such as weather conditions monitoring and performance tracking of equipment in remote areas.

Amateur Radio Networks: Proper introduction of Baofeng radios into computer-based amateur radio networks will increase communication ability and lead to collaboration among radio enthusiasts. Through information transmission protocols like APRS (Automated Packet Reporting System), it becomes easy for individuals to exchange location data, which implies there is instant coordination during field days and other amateur radio activities.

INTERACT WITH THE INTERNET OF THINGS (IOT) PLATFORMS

The following section discusses the integration of Baofeng radios in the IoT sector. These exemplify the flexible uses of Baofeng radios in the projects.

AGRICULTURAL MONITORING SYSTEMS:

The Bofeng radio can function as a data collection relay and transmitter in the monitoring system of agricultural sensors, delivering data from remote field devices to the main unit. For instance, Baofeng radios with sensors for soil moisture levels, temperature, and humidity can be installed for farming fields. The agreement represents the net contribution of data from various sensors toward a central IoT platform for subsequent and effective analysis. It facilitates farmers in fine-tuning import course schedules, monitoring the state of crops, and thus increasing productivity.

ASSET TRACKING APPLICATIONS:

Baofeng radios can be embedded into valuable asset-tracking applications that allow you to monitor the location and status of the valuable item and be in constant touch with it in real-time. For example, various types of handheld radios may be implemented in equipment, vehicles, or inventory items with GPS Modules. Then, these devices can be tracked and monitored to ensure the system's security. IoT platforms that are connected to businesses can track the location of assets in a real-time manner, send notifications for any untoward checkouts, and simplify the supply chain execution processes.

SMART HOME AUTOMATION PROJECTS:

The smart home system delegates the role of Baofeng radios to be a source of communication for IoT devices and systems to interact among themselves. As a case example, Baofeng radio paper kits with wireless modules have the ability to communicate with smart thermostats, lighting systems, and security cameras; thus, they create an interconnected home ecosystem. People can remotely operate and track their intentions at home by using a Baofeng radio as a communication hub to improve their lifestyles. The use of that radio can add to convenience, comfort, and security.

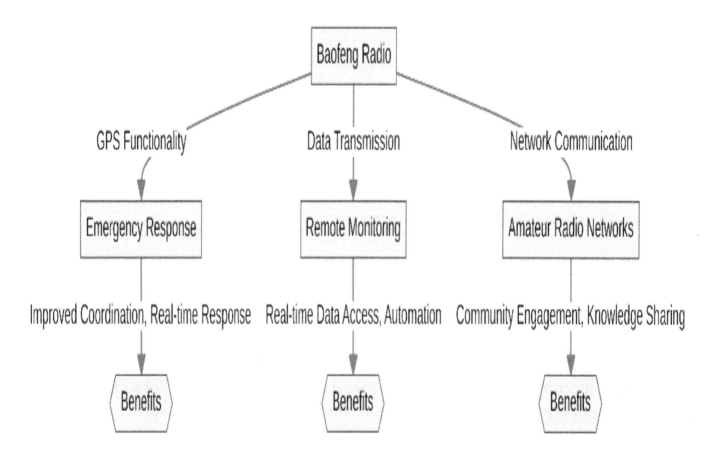

9.2 REMOTE CONTROL APPLICATIONS

Baofeng radios are a vital part of the operation of many applications in places that are not easily accessed. They can operate any unmanned aircraft (UAV) and weather station remotely. Here's a comprehensive overview of some common remote control applications employing Baofeng radios: Here's a comprehensive overview of some common remote control applications employing Baofeng radios:

UAV Operations: Bao Feng radios are primary control system components for connectivity and instant dialog between operators and unmanned aircraft. Through Baofeng radios embedded with UAV systems, pilots in the field can easily and effectively send actions to the UAV, receive live data remote telemetry, and monitor flight condition details without any problem. Our combat accessories enable operators to work with signal channels during UAV operations, providing safety and efficiency in air missions.

Remote Weather Monitoring: Among the main elements of Baofeng weather observations from remotely located weather stations, these radios are crucial for communicating vital meteorological data. These radios are installed in areas that cannot be reached by other long-range methods to relay temperature, humidity, wind speed, and atmospheric pressure parameters. Integrating Baofeng radios and IoT platforms could eliminate the need for manual data collection and transmission. This would enable the collaboration of researchers, meteorologists, and environmental scientists in the monitoring and data analysis of environmental conditions far off. It supports us to know weather patterns better and builds forewarning systems before disasters happen. Also, it helps in researching climatic parameters.

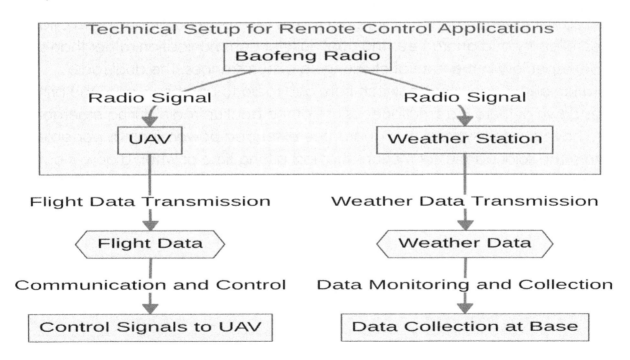

TECHNICAL SETUP AND CONSIDERATIONS

When setting up remote operations using Baofeng radios, several technical considerations should be taken into account to ensure reliable communication and compliance with regulations: When setting up remote operations using Baofeng radios, several technical considerations should be taken into account to ensure reliable communication and compliance with regulations:

Frequency Selection: It's a must to properly comprehend which frequencies are appropriate for long-term operation and also for local requirements and interference. Choose the frequencies used by ham radio and skip crowded frequency bands to diminish the risk of interruption. On top of that, the intensity and location of interference signals in the operating environment should be checked, and frequency allocation should be planned to provide the required communication dependability.

Power Management: The optimum power utilization ensures the battery's longevity and facilitates the smooth functioning of the spacecraft during remote missions when maintenance and human intervention are unavailable. Incorporating power-saving attributes into solutions, including low-power transmission modes and high-efficiency antenna designs, enables the conservation of energy and lengthens the duration of radio operation. Moreover, simultaneously keeping tabs on the battery drains and carrying spare batteries or other substitutes can provide backup power and communication capability even when the mission is faced with emergencies.

Redundancy Planning: Installation of redundant measures is mandatory; otherwise, communication interruption may be possible in far-flung environments. Installing parallel communication systems, such as redundant radios or reserve communication channels, increases reliability and guarantees the continuity of communication rather than simply operation, especially in the face of challenging circumstances. The dual-radio configuration allows operators to switch from one radio to another in case the primary one goes down or there is a simultaneous use of the backup radio during emergencies. They also have support systems that guarantee extended power, such as portable generators and solar panels, for missions that last a long time or when a power outage occurs. For threat mitigation, redundancy planning will allow operators in remote commands to confine the downtime and frequency of failures.

DEVELOPMENTS IN THIS DISTANT CONNECTION

The development of remote communication technologies results in an increased range of features for modern devices' integration synergy with Baofeng. Here's an in-depth exploration of the latest innovations and future possibilities in remote operations: Here's an in-depth exploration of the latest innovations and future possibilities in remote operations:

Drone Swarming: Baofeng radios are pioneers of all sagas of drone cooperation, where different and coordinated missions between thousands of drones have led to the next level with this technology. Nowadays, you can find communication protocols that support radio communication between multiple drones and swarm intelligence algorithms that let them know what to do. This ensures the drones operate as a single unit. Using Baofeng radios, the drone swarm will be able to dynamically allocate tasks, broadcast critical mission data, and respond to different environmental conditions within a matter of seconds. This innovation will result in many applications, including aerial surveillance, search and rescue operations, guided agriculture, and environmental monitoring. Multirotor Swarm Capabilities facilitated by Bayfen Radios would herald a complete penetration into the multiple industrial sectors with unrivaled efficiency, modularity, and scalability in the aerial operations scene.

Satellite Communication: Joining short-range Baofeng radios with satellite communication networks is an important step that gives a chance to extend communication range and increase communication reliability in locations that are hard to reach. Through the LEO satellite networks, Southwind Radio's models can establish global connectivity, creating communication links even in the most isolated regions, oceans, or when aftershocks occur. Along with the satellite communication capabilities, operators no longer have communication problems with their ground teams. They can exchange information with any ground control team and don't have to be in the same field. Whether performing in wilderness areas that are impossible to reach, during missions at sea, or in work that is performed in disasters, Baofeng radios, which use satellite communication technology to keep communications linked strong and working even in circumstances when other components of communication systems would be down, are essential in making communication possible even in situations where other communication tools would be rendered inoperable.

9.3 DATA TRANSMISSION

Baofeng radios possess a wide range of functions and different protocols and data transmission modes, demonstrating their availability in different spheres of activity and data transfer processes. Let's delve into different data transmission methods, security aspects, and case studies highlighting successful projects: Let's delve into different data transmission methods, security aspects, and case studies highlighting successful projects:

METHODS AND PROTOCOLS

Data transmission using Baofeng radios can be achieved through several methods and protocols, each suited to specific communication requirements and operational scenarios Data transmission using Baofeng radios can be achieved through several methods and protocols, each suited to specific communication requirements and operational scenarios:

Voice Communication: However, main data providers risk maintaining voice communication, for example, using Baofeng radios that allow real live exchange of information. Voice has an advantage over chat for communicating precisely yet gets to the point, so it can be used when texting is impractical or inefficient.

Packet Radio: A particular digital communication technique is packet radio, which transmits data digital packets over radio frequencies. Baofeng radios will be integral in accessing digital data, e.g., text messaging, telemetry, and many other information in AX.25 and other protocols using TAPR's AX.25. An aspect that makes the packet radio system stand out from the rest is that it is viable for sending data over long distances and can transmit the data in scenarios with limited structured layout. It can establish ad hoc communication networks.

Frequency Modulation (FM): Bafeng radio brand performance is compatible with FM modulation, which allows the transmission of analog audio signals through the radio frequencies. By and large, our favorite form modulation application is the transformation of the human voice. Yet, other possibilities of digital transmission of signals, such as music and tones, are widespread in applications used by hobby radio amateurs and simplex communications.

Digital Modes: They support digital modes such as DCD, DSD, and P25 of Baofeng Portable Radios. Digital channels deliver higher accuracy, efficiency, and security than their analog counterparts, so they are the best choices for services where reliable and secure data transmission is essential, like public safety, emergency communication, and digital voice networks.

SECURITY ASPECTS:

Encryption: Guided by the encryption abilities of Bahofeng radios, the transmission of messages is secured through the encryption of voice and digital signals by cryptographic algorithms. Encryption sets confidentiality and privacy as the main objectives, allowing undisclosed information to be secure during transmission through a secure channel that avoids the risks of unauthorized access.

Frequency Hopping Spread Spectrum (FHSS): FHSS is a mode of radio modulation used to improve transmission reliability by rapidly random frequency shifts in the predicted spectrum. Baofeng radios, which apply FHSS to foresee the risk of being intercepted and jammed, also can make them better suited for communications in hostile environments or with other contested streams.

CASE STUDIES:

Disaster Response: Designed for a wide variety of disaster scenarios, using Baofeng radios, data transmission during the disaster response is now possible. This facilitates rescue operations coordination, situational awareness data sharing, and real-time real-time communication with the command centers. Through packet radio and digital modes, Baofeng radios take advantage of fail-proof and uninterrupted signals even amid rugged terrains and places where traditional infrastructure might be weakened or altogether lacking.

Wildlife Tracking: Scientists pack the radio with Baofeng radio-electronic equipment, and they use the packet radio capabilities to transmit wildlife movements and behavior in remote wilderness areas. Through the use of radiotracers tagged to the side of wildlife animals and transmitting Baofeng radios as receivers, researchers can collect telemetry data, monitor animal relocation, and learn about habitual utilization patterns, which further support wildlife conservation and management initiatives.

Environmental Monitoring: The deployment of Baofeng radios with environmental sensors in the networking can remotely help monitor environmental parameters such as temperature, humidity, and air quality. Through factual transmission from sensor waives to radio frequencies, Baofeng radios will enable real-time monitoring of the different environmental conditions in remote areas or hard-to-access places where they will be used to provide data support to the researchers and the conservationists of environments and also help in natural resource management.

10. BUILDING YOUR GROUND STATION

The ground station or the base station creates a solid base from which communication can become very reliable and effective. Here's a comprehensive guide on how to construct and optimize your base station:

10.1 DESIGN AND SETUP

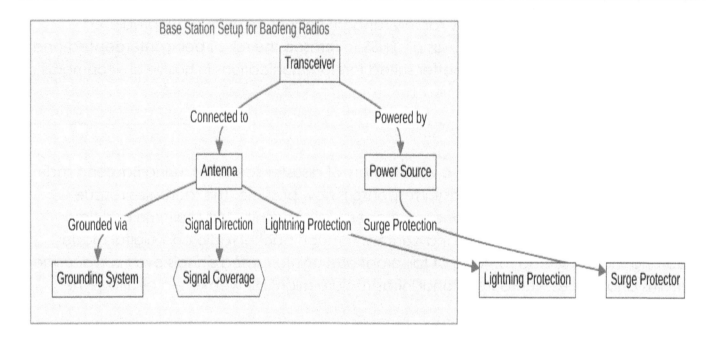

The making of a homestead or a mobile station involves the systematic and attentive installation of building structures or adding more units. Here are detailed steps to help you create a robust communication hub: Here are detailed steps to help you create a robust communication hub:

Choose a Suitable Location: Start with choosing an unobstructed location free of this radio frequency (RF) interference, a source of power that is close by. Preferable places are platforms whose height allows for a long-range view and grounding is even.

Equipment Selection: Provide funds for high-quality transceivers and antennas if you need any accessories that are required for your communication purposes. Review the parameters of frequency range, power output, and durability when choosing the devices.

Antenna Installation: Install the antennas adequately by picking the best location regarding the best height and angle in order to have the best coverage plus minimize interference. Install appropriate mounts, e.g., roof mounts or mast brackets, for the type of antenna and the location it will be set up.

Grounding and Lightning Protection: Protect whatever equipment from electrical surges and lightning strikes using reliable grounding systems and spot-on surge protection. Stabilization rods, current Mats, and lightning arrestors are some of the critical items needed for a properly constructed lightning protection setup.

Equipment Configuration: Tune your transmitter and the accessory unit per the instructions and functional requirements provided by the manufacturer. In order to maintain good performance and maximize power efficiency per unit of energy, tuning of the transmission frequencies, modulation modes, and power levels must be done accurately and in accordance with regulatory standards.

TROUBLESHOOTING AND OPTIMIZATION

Despite the fact that a base station is tightly erected with all its parts, it can have some glitches over time. Here is how to troubleshoot common problems and optimize the performance of your station:Here's a guide on how to troubleshoot common problems and optimize the performance of your station:

Signal Quality Assessment: Simultaneously examine the quality of the direct and indirect signals using signal analyzers and meters. Listen carefully for signal strength, accuracy, and the existence of interference or noise. If the signal is not very good, change the angles of the antenna so that you get better signal reception and transmission.

Antenna Tuning: Tune your antenna regularly to improve its performance constantly. Try to come up with the antenna's length, position, and orientation geared for improving signal propagation. Instruments such as SWR meters should be used to make sure that the connection between the transceiver antenna is optimized and that the unwanted standing wave ratio is lessened.

Equipment Calibration: The transceivers, antennas, and supporting accessories should be calibrated from time to time and adjusted accordingly to ensure that they are precise and working well. The steps to calibrate will also vary according to the specific equipment used, so refer to your supplier's instructions or quick start guide for detailed steps.

System Testing: Performing different ways of system tests and performance evaluations should be a regular habit endowed to the team. Simple measures will help to detect any problems before they escalate. Thoroughly test all the base station components, like transceivers, antennas, cables, and power sources, to confirm they perform as they should without errors. Make sure to take care of all problems and abnormalities quickly to provide the open space facility with the best working performance.

10.2 NETWORKING RADIOS

Establishing a Baofeng network blurs the lines between local and global communication, allowing massive-scale sharing of information. Here's a comprehensive guide on setting up and managing radio networks effectively: Here's a comprehensive guide on setting up and managing radio networks effectively:

CREATING RADIO NETWORKS

To establish and manage networks of Baofeng radios, follow these key steps: To establish and manage networks of Baofeng radios, follow these key steps:

Network Topology Design: Decide on the network architecture that is most appropriate in meeting the needs of communication of your interest. Picking from star, mesh, and point-to-point configurations helps in considering variables like the area of coverage, availability, and redundancy.

Frequency Planning: Channel islands must be alternated on networked radios and, at the same time, share the frequency so that it can be optimally used. While setting up your transmission, it is vital to be aware of local regulations, radio traffic of other transmissions, and possible sources of interference in order to avoid any distortions.

Protocol Selection: A selective network protocol to support efficient communication within your network should be chosen. Choose the mode of operation you require as you weigh factors of range, capacity, and reliability (simplex, duplex, or repeater modes).

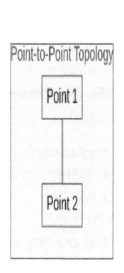
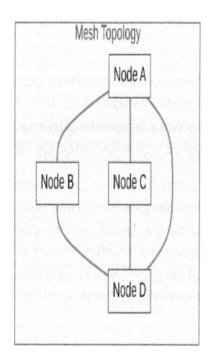
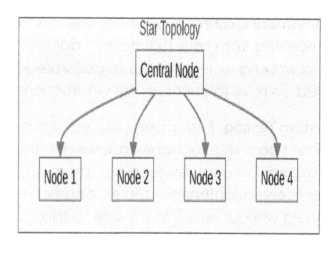

NETWORK MANAGEMENT AND PROTECTION

To ensure the reliability and security of your radio network, follow these best practices: To ensure the reliability and security of your radio network, follow these best practices:

Access Control: Introduce strong access control methods as a way of reducing the probability of unauthorized admissions to your network. Employ encryption and authentication as a protocol for user authentication and encrypting communications to ensure that information with a sensitive nature is not accessed by unauthorized users or intercepted.

Traffic Management: Supervise network throughput and assume high-priority communications to ensure the timeliness of messages and reduce latency. Introduce QoS (Quality of Service) mechanisms to give precedence over voice and emergency signals above data traffic that is not urgent, guaranteeing that significant messages get through the net traffic even in times of high congestion.

Redundancy Planning: Consider designing the layers of your network so as to define the failures other than a single failure of the network. Establish a system with redundant links, backup power sources, and failover capabilities in order to enable uninterrupted operations in the case of equipment problems or a network.

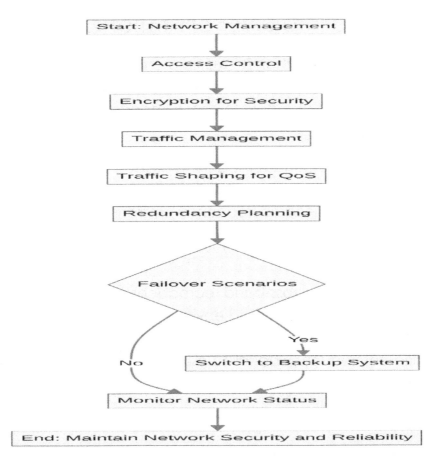

10.3 ADVANCED SCANNING TECHNIQUES

Having the ability to employ filtered and prioritized scanning approaches would empower communication and situational awareness.

DEVELOPING SCANNING SKILLS

Frequency Spectrum Analysis: Use the spectrum analyzers to visualize and interpret power spectra. This enables you to identify busy channels, find interferers, and assess channel selection to get quality communication.

Trunked Radio Systems: Set scanners up to follow trunked radio systems that use the frequency ranges dynamically for user conversation. By following trunked systems, you will hear conversations from across channels, which makes it possible to enjoy conversations without missing out on what's critical.

Legal and Ethical Considerations: Make sure you always follow the approved guidelines for ethical scanning according to the law. Show respect for people's privacy rights and do not get on the path of legal communications. Make sure that you are scanning with proper etiquette and are not disturbing those using the radio for official business. Failures.

EMBEDDING SCANNING INTO DAILY OPERATIONS.

Channel Monitoring: Synchronously monitor multiple channels for prompt related communications and activities within your operational environment. Regularly, it is required to search frequencies to be able to catch important transmissions and have overall situational awareness.

Dynamic Frequency Allocation: Conduct scanning process parameter adjusting constantly according to dynamic operational demands and variable conditions of the environment. Regularly check the radio frequency environment and maintain the scanning efficiency and reactivity. Create AI content like a human would with GPT-4 technology.

Cross-Band Repeating: Utilize the cross-band rebroadcasting features to overcome signal range limitation and to create a networking environment that combines multiple radio networks for enhanced interoperability. Employ cross-band repeaters to distribute messages across different frequency bands to establish fluent communication between various radio sockets.

10.4 INTERACTIVE EXAMPLE

In this part, we will explore dynamic examples explaining the practical use of a communication hub of agents in practical life cases. Picture yourself in a situation where you are handling a common issue or a community event like a marathon or festival where connection with the volunteers is vital for maintaining safety and management logistics.

SCENARIO: COMMUNITY EVENT COMMUNICATION STATIONS.

Assume you are the coordinator of a local marathon. You have numerous electronic devices connected to a dependable communication network that ensures volunteers stationed along the course, medical teams, and emergency responders are communicated promptly. Here's how you can set up a Baofeng communication hub to meet these communication needs: Here's how you can set up a Baofeng communication hub to meet these communication needs:

Step 1: Regarding communication, the key issues have to be analyzed by determining how the stakeholders would be informed about the problem and what tools would be helpful to communicate the issue.

First, determine who will be required to communicate with whom during your occasion and how. Select core participants like event organizers, volunteers, medical professionals, and security teams, and pinpoint their communication requirements. Taking into account the coverage area, type of terrain, expected obstructions, and the number of users should be included.

Step 2: Choose Baofeng Radio Models

Select models from the Baofeng series that provide suitable communication standards for your event. Select different types of transceivers with handy characteristics of being dual-band, having long battery life, and being battery type for toughness outdoors. Budget-friendly devices such as Baofeng UV-5R or UV-82 are good options that can be applied to sporting or commercial-related commercial-related events.

Step 3: Specify the Channels and the Frequencies

After that, the channels and frequencies on the Baofeng radios should be turned on to allow people to communicate even though they are in different parts of the site in the easiest way possible. Classify priority groups and associate them with individual channels, such as race officials, volunteers, medical teams, and security personnel. Guarantee that the frequencies are precise of rules and regulations at that particular locality and compatible with other user areas.

Step 4: Communication Protocols

Put a transparent communication system in place aimed at adequate and fast-flowing information sharing during the event. Develop guides for starting and discontinuing communication via radio, using non-ambiguous signatures and codes with a preference for first-responder messages right away—team event staff and volunteers must have proficient communication skills to ensure a successful event day.

Step 5: Develop and Streamline

It is a must to perform comprehensive testing of the communication system to find all the problems and areas for improvement. Perform range performance tests that show reception and quality in different venue places. Since this is a new product, there could be uncertainties regarding the effectiveness of its channels, frequencies, and communication protocols; thus, when performing tests or receiving user feedback, these areas should be assessed and adjusted accordingly.

Following these steps, you can establish a Baofeng radio communication center for the community festival, which will favor the quick flow of information and coordination during the event gathering. At any event where we're bringing people together, from a marathon or a festival to other community gatherings, a skillful communication system is vital because it is the backbone of the event's optimal coordination and safety.

10.5 DIGITAL INTEGRATION

Whether it is a computer program for managing radio networks or an interface for digital performance monitoring, this is the spectrum of the solutions we will look into to get the best out of your communication system. Whether you're talking about digital voice modes, repeater controllers, or network monitoring tools, this article aims to convey informative knowledge and practical wisdom in a way that will smoothly combine them for your success.

DIGITAL VOICE MODES

Digital voice modes (such as Digital Mobile Radio (DMR) and D-STAR) are part of the recent development in radio communication technology. Those digital methods stand out with remarkable features such as the unmatched quality of speech, clear signals, and dozens of eavesdropped terminals instead of one. Using DMR/D-STAR with your Baofeng radio hub, you can get benefits here and increase the communication means of your team simultaneously.

REPEATER CONTROLLERS

Another area where digital technologies are applied is replacing the older units with repeater controllers, thus increasing the performance of repeater systems. Repeater controllers are pieces of equipment that solve the problem of managing the operation of repeaters, which are devices used for controlling access, coordinating transmissions, and providing many additional features, such as voice announcements and status monitoring). With the repeater kit controller, you can push the range and the coverage of your communication net, and you will perform better by providing advanced features that will enhance usability and efficiency.

NETWORK MONITORING TOOLS

To guarantee the validity and productivity of your communications center, it is pivotal to keep track of the status and output of your radio network. The network monitoring tools help technical personnel track essential metrics like signal strength, packet loss, and network latency while providing insights into the communication structure. By incorporating network monitoring tools in your Baofeng communication hub, you'll be able to detect and resolve problems beforehand, achieve maximum network performance, and thus maintain your communication uninterrupted during periodic activities.

SOFTWARE SOLUTIONS

Another factor in radio network handling is the use of software in addition to the hardware, which is also an essential tool in managing and optimizing radio networks. There is an extensive library of software that provides

- Solutions for programming Baofeng radios,
- Managing the configurations of the channels and
- Analysis of communication information.

Using software specific to your need, you can bring together all the administrative tasks, the general repetitive processes, and the overall productivity of your hub.

10.6 SECURITY CONSIDERATIONS

Security has to be an issue for those concerned about the development of this communication center. Hence, I will emphasize the challenges of cyber threats to radio networks. Our attention will be dedicated to developing tactics to reduce risks and, at the same time, to network security and integrity. The subjects we cover deal with, among others, encryption methods and access control mechanisms, network segmentation, and secure authentication techniques, which will bestow enough theoretical and practical knowledge to defend your communications hub from unauthorized access and risks.

ENCRYPTION PROTOCOLS

Performing a competent Encryption Protocol is crucial to the safety of the communication data, as it may be accessed or intercepted by unauthorized persons. You'll get to know encryption standards such as AES (Advanced Encryption Standard) and DES (Data Encryption Standard), which we'll analyze to shed light on their pros and cons and provide the choice of the most suitable encryption technique for your communication hub.

ACCESS CONTROL MECHANISMS

Establishing solid access controls is essential to preventing interception by an intruder from accessing your communication network. We will give you guidelines on different techniques, including password authentication, biometric authentication, and RBAC, and we will offer you some hints that can help implement the correct access control policies that meet your security requirements.

NETWORK SEGMENTATION

Seeing off your information net efficiently by forming distinct, constitutional zones beside one another may restrict the spread of security spills and the spread of assaults. The topic of network segmentation will be introduced, and guidance on implementing zones that match the safe boundaries of the network will be given. Here, user roles, device types, and communication protocols will play a role in deciding the compromises related to creating the network segments.

SECURE AUTHENTICATION METHODS

Criminal threats are lower when robust authentication methods ensure only the right users and devices can access their communication hub. Authenticity through MFA, digital certificates, and cryptographic keys will be rooted here. We shall highlight their weaknesses and strengths so you can apply them appropriately after implementing them.

CONCLUSION

Looking at the future, the ability to develop and master these skills will be pivotal in securing one's success.

With the final chapter of radio communication being written, it is safe to say that the environmental learning and advancing field of radio communication cannot be overrated. Radio communication tech schemes' texture and usage patterns are evolving continuously due to ongoing innovations and shifts in regulations and usage purposes. To be ready for the changing world in communication skills, you need to adopt the attitude of Lifelong Learning and Adaptation.

CONTINUOUS LEARNING

Radio communication is in a process of swift transition. In this regard, being aware of the recent developments, techniques, and technologies is vital. Whether you are a radio hobbyist or just using the radio as your gadget, constant education and development of new skills allow you to keep pace with the times.

PREDICTING FUTURE TRENDS

Finally, some predictions are based on the fact that the personal technology of radio communication and practice will change over the years:

Convergence of Technologies: The use of radio with the latest technical innovations like artificial intelligence, the Internet of Things, and augmented reality will result in new applications and capabilities that will dictate the kind of communication and interaction we have with our environment.

Enhanced Connectivity: With the expansion of high-speed internet connectivity and the completion of 5G networks, those of us who have personal radio communication devices will enjoy better interconnection and become fully interoperable, leading to a flow in communications across devices and platforms.

Focus on Sustainability: With the increasing environmental problems, there will be greater attention to researching the development of environmentally friendly and green communication devices. This might involve the use of technologies that are energy efficient, the development of products made of recyclable materials, and manufacturing processes that are environment friendly.

Security and Privacy: As the severity of cyberattacks and privacy breaches rises, greater concerns will be observed around the promotion of security as well as privacy in these systems. For this, developing strong encryption protocols, creating reliable authentication methods, and tough privacy criteria are important.

GLOSSARY

Here's a glossary of terms related to Baofeng radios and radio communication:

Antenna: A device used to transmit and receive radio frequency signals.

Frequency: The number of cycles per second of a radio wave, measured in Hertz (Hz).

Transceiver: A device that both transmits and receives radio signals.

VHF: Very High Frequency, a radio frequency range from 30 MHz to 300 MHz.

UHF: Ultra High Frequency, a radio frequency range from 300 MHz to 3 GHz.

Dual-band: Capable of operating on two different frequency bands.

SWR: Standing Wave Ratio, a measure of impedance matching between a radio transmitter and its antenna.

RSSI: Received Signal Strength Indicator, a measure of the power level of a received radio signal.

CHIRP: Comms Help In Radio Programming, a software tool for programming Baofeng radios.

Base station: A stationary radio unit used for communication within a localized area.

Mobile radio: A radio unit designed for installation in vehicles for mobile communication.

Handheld radio: A portable radio unit designed for handheld use.

Duplex: A communication mode that allows simultaneous transmission and reception on different frequencies.

Simplex: A communication mode that uses the same frequency for both transmission and reception.

Repeater: A device that receives a radio signal on one frequency and retransmits it on another frequency to extend the communication range.

CTCSS: Continuous Tone-Coded Squelch System, a method of sub-audible tone signaling used to mute the receiver until a matching tone is received.

DCS: Digital-Coded Squelch, a digital signaling method used to control access to a shared radio channel.

VOX: Voice-Operated Transmission, a feature that automatically activates the transmitter when the user speaks into the microphone.

Squelch: A circuit that mutes the audio output of a receiver when no signal is present.

APRS: Automatic Packet Reporting System, a digital communication protocol used for real-time tactical communication and tracking of objects.

GPS: Global Positioning System, a satellite-based navigation system that provides location and time information.

Encryption: The process of encoding information to prevent unauthorized access.

Firmware: Software that is embedded into hardware devices to control their operation.

Programming cable: A cable used to connect a radio to a computer for programming purposes.

Programming software: Software used to program the settings and configurations of a radio.

Battery: A device used to store and supply electrical energy to power a radio.

Charger: A device used to recharge the battery of a radio.

Solar charger: A charger that uses solar panels to convert sunlight into electrical energy for recharging batteries.

Desktop charger: A charger designed to recharge multiple radio batteries simultaneously.

Belt clip: A clip attached to a radio for securing it to a belt or clothing.

Holster: A protective case or pouch for carrying a radio on a belt or strap.

Protective case: A case designed to protect a radio from damage, dust, and moisture.

Tactical antenna: An antenna designed for rugged outdoor use that is typically made of durable materials.

High-gain antenna: An antenna designed to amplify radio signals for increased range.

Dual-band antenna: An antenna designed to operate on two different frequency bands.

Speaker microphone: A microphone with a built-in speaker for hands-free communication.

Surveillance earpiece: An earpiece used for discreet communication, typically used in security applications.

Bluetooth headset: A wireless headset that connects to a radio via Bluetooth technology.

Squelch level: The threshold at which the squelch circuit opens to allow received signals to be heard.

Repeater offset: The frequency difference between the input and output frequencies of a repeater.

Tone mode: A setting that determines whether a radio uses CTCSS or DCS tone signaling.

Scan mode: A mode that allows a radio to search for and monitor active channels automatically.

Priority channel: A channel that is scanned more frequently than other channels.

Weather alert: An alert function that automatically switches to weather channels when severe weather warnings are issued.

Call tone: A tone used to alert other users on the same channel.

Dual watch: A feature that allows a radio to monitor two channels simultaneously.

Frequency range: The range of frequencies that a radio is capable of transmitting and receiving.

Rechargeable battery: A battery that can be recharged and reused multiple times.

Non-rechargeable battery: A battery that cannot be recharged and must be replaced when depleted.

Memory channel: A programmable channel used to store frequently used frequencies and settings.

FM: Frequency Modulation, a method of encoding information onto a radio wave by varying its frequency.

AM: Amplitude Modulation, a method of encoding information onto a radio wave by varying its amplitude.

S-meter: A signal meter is a display indicating the strength of received signals.

Channel spacing: The frequency separation between adjacent channels.

Power output: The amount of power that a radio transmitter is capable of producing.

Receiver sensitivity: The minimum signal strength that a radio receiver can detect.

Menu mode: A mode that allows the user to access and configure advanced settings and features.

Keypad lock: A function that prevents accidental changes to the radio settings by locking the keypad.

Voice prompt: An audible announcement or indication of radio functions and settings.

VOX sensitivity: The threshold at which the VOX feature activates the transmitter.

Priority scan: A scan mode that prioritizes certain channels over others.

Dual PTT: Dual Push-To-Talk, a feature that allows the user to transmit on two different channels simultaneously.

DTMF: Dual-Tone Multi-Frequency, a method of encoding numbers and symbols for signaling and control purposes.

Battery save mode: A power-saving feature that reduces the radio's power consumption to extend battery life.

Low battery alert: An alert function that notifies the user when the battery voltage drops below a certain threshold.

Time-out timer: A feature that limits the duration of each transmission to prevent the radio from being tied up indefinitely.

Reverse frequency: A feature that allows the user to transmit and receive on the reverse frequency of a repeater.

Dual display: A feature that allows the user to monitor two different frequencies simultaneously.

FM radio: A feature that allows the user to listen to FM radio stations on the radio receiver.

Bandwidth: The range of frequencies occupied by a radio signal.

PTT: Push-To-Talk, a button or switch used to activate the transmitter for voice transmission.

Noise reduction: A feature that reduces background noise and interference during reception.

Timeout beep: An audible indication that the time-out timer has been activated during transmission.

Emergency alert: A function that sends out an emergency signal when activated.

Auto power-off: A feature that automatically turns off the radio after a period of inactivity to conserve battery power.

Lockout: A function that prevents certain channels or frequencies from being scanned.

Automatic squelch: A feature that automatically adjusts the squelch level based on the strength of the received signal.

Memory backup: A feature that stores radio settings and configurations in non-volatile memory to prevent data loss.

Scan resume: A feature that resumes scanning after a transmission is completed.

Transmit power: The amount of power that a radio transmitter outputs during transmission.

Data port: A port used to connect external devices or accessories to the radio for data exchange.

Transmit inhibit: A feature that prevents the radio from transmitting while scanning or receiving signals.

Tone burst: A brief burst of tone used to open repeater squelch or activate remote functions.

Emergency button: A dedicated button used to activate emergency alerts or functions.

Automatic mode selection: A feature that automatically selects the operating mode (FM, AM, etc.) based on the frequency being used.

Memory group: A group of memory channels that can be scanned or accessed together.

Busy channel lockout: A feature that prevents the radio from transmitting on busy channels to avoid interference.

Repeater directory: A database of repeater frequencies and locations stored in the radio for quick access.

Priority watch: A feature that monitors a specific channel for activity while scanning other channels.

Weather scan: A scan mode that automatically searches for and monitors weather channels for alerts and updates.

Digital mode: A mode of operation that uses digital modulation techniques for transmitting and receiving data.

Analog mode: A mode of operation that uses analog modulation techniques for transmitting and receiving voice or data.

Narrowband: A transmission mode that uses a narrower bandwidth than standard FM for improved spectral efficiency.

Wideband: A transmission mode that uses a wider bandwidth than standard FM for higher audio fidelity.

DMR: Digital Mobile Radio, a digital radio communication standard used for professional and commercial applications.

D-star: Digital Smart Technologies for Amateur Radio, a digital voice and data protocol used in amateur radio.

Fusion: Yaesu System Fusion is a digital communication system for amateur radio.

P25: Project 25, a suite of standards for digital radio communication used by public safety agencies and first responders.

NXDN: Next Generation Digital Narrowband is a digital radio protocol used in professional mobile radio systems.

TDMA: Time Division Multiple Access, a digital transmission technique that divides radio channels into time slots for multiple users.

DCS code: Digital-Coded Squelch code, a series of digital tones used to control access to a shared radio channel.

Subaudible tone: A tone used to transmit data or control signals below the audible frequency range.

Prepper: An individual who actively prepares for emergencies or disasters, often including radio communication in their preparations.

Ham radio: Amateur radio is a hobby and service that allows licensed individuals to operate radio equipment for non-commercial purposes.

DXing: Amateur radio operators' practice of communicating over long distances, often internationally.

QSL card: A postcard-sized card exchanged between amateur radio operators to confirm two-way radio contacts.

Net: A scheduled meeting or gathering of amateur radio operators on a specific frequency and time for communication.

Ragchew: Casual and extended conversation between amateur radio operators, often on topics unrelated to radio.

Elmer: An experienced and knowledgeable amateur radio operator who mentors and assists newcomers to the hobby.

Contesting: A competitive event in amateur radio where operators attempt to make as many contacts as possible within a specified time frame.

DXpedition: An organized trip or expedition by amateur radio operators to a remote or rare location for the purpose of making long-distance contacts.

Satellite communication: Communication via amateur radio satellites orbiting the Earth.

Packet radio: A digital communication mode in amateur radio that sends data packets over radio waves.

MARS: Military Auxiliary Radio System, a program that provides auxiliary communication support to the U.S. Department of Defense.

RACES: Radio Amateur Civil Emergency Service, a program that provides emergency communication support to civil authorities during times of crisis.

ARES: Amateur Radio Emergency Service, a program that provides emergency communication support to communities during disasters and emergencies.

SKYWARN: A program that trains amateur radio operators and weather enthusiasts to report severe weather conditions to the National Weather Service.

Public service event: A non-profit event, such as a charity race or parade, where amateur radio operators provide communication support.

Ground plane antenna: An antenna that uses a flat conducting surface as a ground plane to improve radiation efficiency.

Yagi antenna: A directional antenna consisting of multiple elements arranged in a line, commonly used for long-distance communication.

Dipole antenna: A simple antenna consisting of two conductive elements connected to a transmission line.

J-pole antenna: A type of end-fed dipole antenna that uses a quarter-wavelength matching section to improve performance.

Slim Jim antenna: A type of end-fed dipole antenna with a matching section that allows it to be fed with coaxial cable.

Beam antenna: A directional antenna that focuses radio energy in one direction, typically used for long-distance communication.

Vertical antenna: An antenna that radiates radio energy in a vertical pattern, commonly used for mobile and base station installations.

Grounding: The process of connecting an electrical circuit or device to the Earth to prevent electrical shock and interference.

Lightning protection: Measures taken to protect electronic equipment from damage caused by lightning strikes.

SWR meter: A device used to measure the standing wave ratio (SWR) of an antenna system.

Coaxial cable: A type of electrical cable consisting of an inner conductor surrounded by a tubular insulating layer and a tubular conducting shield.

Balun: A device used to balance the impedance of a balanced transmission line to an unbalanced load, such as an antenna.

Dummy load: A device used to simulate an antenna for testing and tuning transmitters without radiating radio waves.

RF amplifier: A device used to increase the power of a radio frequency signal.

Attenuator: A device used to reduce the strength of a radio frequency signal.

Bandpass filter: A device used to pass signals within a certain frequency range while attenuating signals outside that range.

Notch filter: A device used to suppress signals within a specific frequency range while passing signals outside that range.

SWL: A shortwave listener is an individual who listens to shortwave radio broadcasts for entertainment or information.

QRM: Man-made interference on a radio signal, often caused by electrical devices or other radio transmissions.

QRN: Natural interference on a radio signal, such as static or atmospheric noise.

QRZ: An amateur radio operator's callsign, often used as a greeting or identifier.

CQ: A general call used by amateur radio operators to invite other operators to make contact.

QRP: Low power operation in amateur radio, typically using 5 watts of power or less.

QRO: High power operation in amateur radio, typically using more than 100 watts of power.

RST: Readability, Signal Strength, and Tone, a system used to report signal quality in amateur radio contacts.

SWL station: A listening station equipped with a receiver for monitoring shortwave radio broadcasts.

QTH: Location or position, often used by amateur radio operators to refer to their operating location.

QSL bureau: A service that facilitates the exchange of QSL cards between amateur radio operators.

QTH locator: A system used to specify locations on the Earth's surface using a grid of alphanumeric codes.

QSY: Change frequency, often used in amateur radio, to indicate a change in operating frequency.

QRPp: Ultra-low power operation in amateur radio, typically using less than 1 watt of power.

SWL club: A group or organization of shortwave listeners who share information and resources related to shortwave radio.

QRP club: A group or organization of amateur radio operators who specialize in low-power operation.

Hamfest: An amateur radio convention or gathering where enthusiasts buy, sell, and exchange equipment and information.

HAM radio license: A legal authorization granted by a regulatory authority that permits individuals to operate amateur radio equipment.

Technician class license: The entry-level amateur radio license in the United States, granting operating privileges on certain frequency bands.

General class license: An intermediate-level amateur radio license in the United States granting expanded operating privileges.

Amateur Extra class license: The highest level amateur radio license in the United States, granting full operating privileges.

Call sign: A unique identifier assigned to an amateur radio operator or station.

Vanity call sign: A custom call sign chosen by an amateur radio operator, typically based on personal preference or significance.

Morse code: A method of transmitting text characters as a series of on-off tones, clicks, or lights.

Straight key: A manual telegraph key used to send Morse code by hand.

Paddle key: A type of telegraph key that uses two paddles to create Morse code characters.

Bug key: A type of telegraph key that uses a mechanical oscillator to create Morse code characters automatically.

Morse code proficiency: The ability to send and receive Morse code characters accurately and efficiently.

Q-code: A set of abbreviations used in amateur radio communication to convey information concisely.

CW: Continuous Wave, a transmission mode used for sending Morse code signals.

QRP transmitter: A low-power radio transmitter used for QRP (low power) operation in amateur radio.

QRO amplifier: A high-power radio amplifier used for QRO (high power) operation in amateur radio.

RF grounding: A method of grounding a radio antenna to minimize the risk of electrical shock and interference.

RF choke: A device used to block the flow of radio frequency currents while allowing direct current to pass.

SWR bridge: A device used to measure the standing wave ratio (SWR) of an antenna system.

SWR analyzer: A device used to measure and analyze the standing wave ratio (SWR) of an antenna system.

RF filter: A device used to block unwanted radio frequency signals while allowing desired signals to pass.

RF gain: A control on a radio receiver that adjusts the amplification of the received signal.

RF preamplifier: A device used to amplify weak radio frequency signals before they are received by a radio receiver.

RF mixer: A device used to combine two or more radio frequency signals into a single output signal.

RF modulator: A device used to modulate a radio frequency carrier wave with an audio or video signal.

RF demodulator: A device used to extract the audio or video signal from a modulated radio frequency carrier wave.

RF attenuator: A device used to reduce the strength of a radio frequency signal without distorting its waveform.

RF oscillator: A device used to generate radio frequency signals for transmission or local oscillation.

RF amplifier: A device used to amplify radio frequency signals for transmission or reception.

RF switch: A device used to route radio frequency signals between different input and output ports.

RF power meter: A device used to measure the power of a radio frequency signal.

RF connector: A device used to connect coaxial cables or other transmission lines to radio frequency equipment.

RF balun: A device used to balance the impedance of a balanced transmission line to an unbalanced load, such as an antenna.

RF probe: A device used to measure radio frequency signals without making direct electrical contact.

RF coupler: A device used to couple radio frequency signals between two transmission lines.

RF transformer: A device used to match the impedance of one circuit to another circuit at radio frequencies.

RF dummy load: A device used to simulate an antenna for testing and tuning transmitters without radiating radio waves.

RF isolator: A device used to prevent unwanted radio frequency signals from entering or leaving a circuit.

RF splitter: A device used to split a single radio frequency signal into multiple output signals.

RF combiner: A device used to combine multiple radio frequency signals into a single output signal.

RF circulator: A device used to route radio frequency signals in a specific direction within a circuit.

RF filter: A device used to block unwanted radio frequency signals while allowing desired signals to pass.

RF transmitter: A device used to transmit radio frequency signals over the air.

RF receiver: A device used to receive radio frequency signals from the air.

RF antenna: A device used to transmit and receive radio frequency signals wirelessly.

RF cable: A cable used to connect radio frequency equipment to antennas, amplifiers, or other devices.

RF shielding: A method of enclosing radio frequency equipment in a shielded enclosure to prevent electromagnetic interference.

RF spectrum: The range of radio frequencies used for communication, typically from a few kilohertz to several gigahertz.

RF interference: Electromagnetic interference caused by unwanted radio frequency signals.

REFERENCES

BaoFeng Tech. (2023, November 10). About BaoFeng. https://baofengtech.com/

Buy Two Way Radios. (n.d.). BaoFeng Radios. https://baofengtech.com/shop/

FCC. (2023, December 8). Amateur Radio Service. https://www.fcc.gov/wireless/bureau-divisions/mobility-division/amateur-radio-service

My BaoFeng. (2021, March 1). Getting Started with BaoFeng Radios. https://m.youtube.com/watch?v=0wCLbiHvsMg

Radio Reference. (n.d.). BaoFeng UV-5R. https://forums.radioreference.com/tags/baofeng-uv-5r/

The BaoFeng Review. (2022, August 10). A Beginner's Guide to BaoFeng Radios. https://www.amazon.com/Baofeng-UV-5R-136-174-400-480Mhz-1800mAh/product-reviews/B074XPB313

RepeaterBook. (n.d.). Find a Repeater. https://www.repeaterbook.com/index.php/en-us/

The ARRL. (n.d.). Getting Started with Amateur Radio. https://www.arrl.org/

Book References:

Carr, K., & Kennon, M. (2020). The Complete Guide to BaoFeng Radios: Programming, Usage, Modifications. https://www.amazon.com/Baofeng-Radio-Bible-Step-Step/dp/B0CPHG2K7H 1

Gervasi, M. S. (2018). BaoFeng UV-5R User's Guide and Programming Reference. https://www.amazon.com/Baofeng-Radio-Handbook-Beginners-troubleshooting/dp/B0CCCXMWFJ

Ham Radio for Beginners. (2021). BaoFeng Radios for Beginners. [Book]

Hobbysz, M. (2017). Scanner & Amateur Radio Frequency Guide. [Book]

Johnson, G. (2022). BaoFeng Radios: A Practical Guide to Using and Programming Two-Way Radios. [Book]

Kelly, M. (2019). The BaoFeng Handbook: A Beginner's Guide to Programming and Using Your BaoFeng Radio. [Book]

Kikut, R. (2020). BaoFeng UV-5R User's Guide: Programming and Modifications. [Book]

Lindenborg, H. (2018). The BaoFeng Airband Scanner Manual. [Book]

McGreevy, T. (2021). BaoFeng Radios for Dummies. [Book]

Scherer, A. (2019). The BaoFeng UV-B5 User's Guide: Programming and Modifications. [Book]

Smith, C. (2022). BaoFeng UV-82 User's Guide: Programming and Modifications. [Book]

Stone, G. (2020). The Complete BaoFeng UV-5R Programming Guide. [Book]

Made in the USA
Coppell, TX
11 October 2024

38520082R00070